Liberated Time

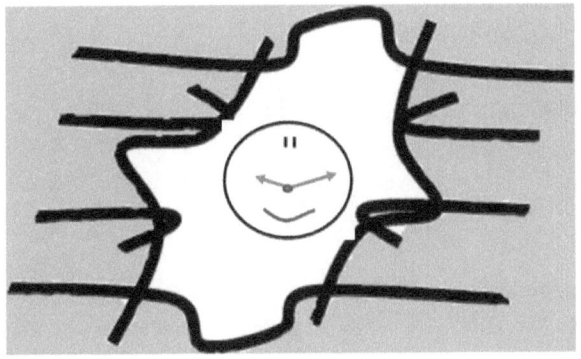

Liberated Time

NEW APPROACH TO MICHELSON-MORLEY EXPERIMENT AND
HOW IT RESHAPES OUR UNDERSTANDING OF
MICRO– AND MACROWORLDS

Val Parker

Copyright © 2018 by Val Parker.

Library of Congress Control Number:		2018900914
ISBN:	Hardcover	978-1-5434-8009-2
	Softcover	978-1-5434-8010-8
	eBook	978-1-5434-8011-5

All rights reserved. No part of this book may be reproduced or transmitted in any form or by any means, electronic or mechanical, including photocopying, recording, or by any information storage and retrieval system, without permission in writing from the copyright owner.

Any people depicted in stock imagery provided by Getty Images are models, and such images are being used for illustrative purposes only. Certain stock imagery © Getty Images.

Print information available on the last page.

Rev. date: 03/26/2018

To order additional copies of this book, contact:
Xlibris
1-888-795-4274
www.Xlibris.com
Orders@Xlibris.com
770825

CONTENTS

Acknowledgements vii
PREFACE ix
ABOUT THE BOOK... xi
HOW THE BOOK IS ORGANIZED xiii
INTRODUCTION... xv
A Legacy of Einstein xix

PART I
AT THE CRADLE OF ELECTRODYNAMICS

FROM JAMES CLARK MAXWELL TO ALBERT
    EINSTEIN.. 3
THE WORLD OF FUNNY MODELS AND WITTY
    FANTASIES.. 9
First Step toward the "Dark Side" 9
Into the "Dark Side"... 15
On the Dark Side... 21

PART II
AT THE SOURCE OF EVERYTHING

From the Darkness into the Light 33
E-Model 38
The Theory of Dual Transformation.. 52
The First Transformation 53
Second Transformation 55
Gravity as the Energy Absorption Process............................. 57
Is a quantum of energy a particle?.. 58
Quantizing in Electrodynamics 60

PART III
NEW MODEL AT WORK

The Gravitational Model at Work 63
Characterizing Space and Time 65
Space 66
Time 67
The Big Bang Theory 70
The Process of Sun Burning 74
Description of Drastic Changes of the Earth's Climate 76
Black Hole Theory in the View of Our Exchange Energy
    Model 78
The "Parker Effect" 80
From a Light Box to the Velometer 88
The Velometer—the Device of Noninertial Navigation 92
Space Telescopes or "What We Have Not Learned from
    Troubles with the Hubble" 93
The "Parker Effect" and the Optical Alignment of the
    Space-Based Telescopes 95
The Velometer Device as a Planetary Compass 100
Contemporary Relativity 102

PART IV
ON KINETIC AND POTENTIAL ENERGIES

The Contemporary Views on Kinetic and Potential Energies ... 107
AFTERWORD 117
REFERENCES 119

Acknowledgements

I would like to express my deep gratitude to my wife for making this book readable and understandable for general readers. She undertook this tumultuous task against her belief that it is undeniably impossible to change our mind-set of the current deeply entrenched system of science.

PREFACE

For more than a century, people were intrigued by the discoveries made at the end of the nineteenth and beginning of the twentieth centuries. Works on electromagnetism, electrodynamics, and nuclear physics changed our understanding of the Universe. Technological breakthroughs open doors for exploration of space and the digital revolution. On the other hand, current interpretations of Nature in scientific communities still insist on a "zero time," a "starting point," and the beginning of the Universe itself. Scientists continue their search for gravitational waves of Einstein's General Relativity, multidimensional space/time mysteries, inexplicable dark matter, etc. All that were underlined by the centuries-old attraction to outdated interpretations of all energy forms, including Kinetic and Potential energies, which were even utilized in quantum physics.

This book is my highly individual approach to theoretical and applied physics, which resulted from decades of searching for answers to the most common and intriguing questions.

Among those was a question, why does not the famous $E = mc^2$ equation ascribe energy to a photon even though the energy of photons is a well-known and recorded fact?

The views expressed in this book are obviously all my own, and all are based on the knowledge I have gained during my studies and while searching for answers in the works of many authors. The references that played a major part in formulating and clarifying my ideas are highlighted in Italics in the reference section. My deepest gratitude to my physics teacher Nathan Guberman, who has profound understanding of physics, who was and continues to be my lifelong inspiration.

If I will succeed in even marginally altering traditional understanding of the phenomenon of Nature and generate more radical stratum of independent intellectuals, I will consider this book worthwhile.

ABOUT THE BOOK

This unique book contains the keys toward understanding the entire Universe. Science is a collection of the understanding of the ways of Nature, and it simply reflects the current state of our understanding of such. The book reflects my commonsense view of Nature, as well as my attempt to bring into order the chaos of the current quilt of stitched-together theories.

It starts with the searching for the answer to the question:

Which Einstein is right?

Is it the young Einstein, author of *Special Relativity*, or the Mature Einstein, mastermind of *General Relativity*?

There has since been no middle ground to have both works to be right.

The book takes the reader on the trail to search for the solution to the conundrum of the Michelson-Morley experiment. The results of this search lead to the discovery of the unique entity called "Prime Energy" and how it is responsible for the existence of everything in the Universe. Furthermore, in the new light of this discovery, the reader will come across many interesting facts about the Big Bang, Black Hole, Earth Climate, Hot Burning Suns, and other details of Nature.

The book contains descriptions of the interesting technical discovery of the "Velometer" -- the device that provides 3-D spatial information on an object's position and velocity from within and from the motion itself. This discovery will guide astronauts deep into the depth of the Universe and safely bring them back Home. On Earth, Velometer-type devices will help manned and unmanned craft to travel safely and Autonomously.

The book will be equally interesting for professional Scientists as well as for recreational readers.

HOW THE BOOK IS ORGANIZED

Main Entries
Regular continuous entries by the author are in regular font.
Theoretical conclusions and important remarks by the author are placed inside text boxes.
Entries from references or outer sources are made in accordance with the chapter's subject and placed inside quotes, using Italic font.

General Outlines
The book consists of four parts that are numbered from Part 1 through Part 4.
Parts have chapters that are not numbered but separated by the Bold Headings.
Some chapters have subchapters that are numbered and separated by the Bold Headings.

Equations
Equations are identified by the numbers of the Part where they are located within.
Notable equations are typed in Bold font.
Some equations are listed in the traditional symbol forms.

To Strive, to Seek, to Find, and Not to Yield

INTRODUCTION

The end of the nineteenth Century and the beginning of the twentieth Century were characterized by winds of freedom, which were driving revolutions in every aspect of human life, including natural science. There were times when "Time" was liberated from the eternal chains that held it affixed to nonexistence - "Beginning of Time." It is well-known that it was not Prometheus, Zeus, or Hercules but young Einstein who dared to cut the chains and let "Time" be free.

It was a short and powerful burst of winds of freedom that quickly subsided, and the man who liberated "Time" afterward single-handedly locked it up behind the bars of "Space-Time" continuum. One hundred years of the incarceration of "Time" has passed, and it is up to us to undo injustice and liberate "Time" once and for all. I truly hope that the work described in this book will initiate the Liberation of Time from the restraints of the Space-Time continuum.

Looking back, I was fortunate in post-WWII and after Stalin's GULAG era to have teachers and professors who learned from association with world-renowned physicists like Paul Dirac and Lev Landau. I was even more fortunate when in mid-1980, the Iron Curtain was raised a bit and my family, and I were able to squeeze through this little opening into the freedom in the USA. It is impossible, without involving providence, to explain my unbelievable luck of discussing some of my ideas with Hans Bethe and Edward Teller - the two great minds of the twentieth century and the contemporaries in the great scientific revolution.

I believe that now it is my turn and my obligation to pay back for my fortune of learning from the best minds, and I hope to accomplish that in this work.

My intention is to provide a clear and unbiased picture of general science and, most especially, Physics from the perspective of the individuals who witnessed prominent changes that took place in the scientific community over the last half of the century. This is the time frame in which the strong lure of excitement of a discovery generated at the first half of the twentieth century was replaced with a swamp-like existence of comfortable chairs and lavish titles in its second half. As it frequently transpires in life, obvious potentials created by explorers and risk-takers became attractive nesting grounds for all sorts of bureaucrats who acclaimed the field for their own good.

Past centuries brought us great discoveries that covered every part of Nature from the Macroworld of Astronomy to the Microworld of a Nucleus. Great minds of our past left us with lavish inheritance of scientific knowledge that is needed to be analyzed, utilized, and expanded in the ways that someday it will become such a worthy inheritance of the future generations.

History taught us that science is not built on dogmas, doctrines, and convictions of an individual, regardless of his/her standing in society. Visions that were created by the knowledge of yesterdays can be refocused or rather replaced by the reality of the present. Our goal of today is to intelligently sift through scientific material that we have inherited from the past, using the present state of understanding that we have built on the foundation of knowledge that was developed by previous generations. To establish actual criteria in science, we must research the very spring from which a stream of knowledge is flowing. Only after that can we be trustful that by trailing this stream, we may achieve the desired results and get to the foreseen destination.

In this work, I have followed the stream of knowledge that sprung from Maxwell's theoretical works, Michelson-Morley experiments, and right into the waters of Special and General Relativity, Energies of Macroworlds and Quantum worlds. Our destination would be the territory of the E-field, Energy Exchange, and Dual Transformation of Energy, which are the sources of our existence.

At this point, I would like to welcome you aboard this journey into the known, unknown, and partially forgotten world of the science of Nature.

A Legacy of Einstein

Since Einstein's death, he is continuously cast as a lifeless Monument for the sake of selfish gains and mercantile profiteering. This transformation striped the human quality of the Man, providing a chance to speak on his behalf, rewrite his statements, and reinterpret his work and the nature of his accomplishments. This was made possible because anyone may hug the monument and say or write "Einstein and I . . ." - hence putting himself/herself on equal fitting. Once I came across an article where the author shamelessly wrote "As Dr. Einstein stipulated . . ." and then signed the article "Dr. so-and-so."

As a result, significant number of the current interpretations of Einstein's work are skewed to such a degree that they contradict the very essence of the original.

The goal of this chapter in the Book is to go back to the beginning of the twentieth Century, to observe Einstein's work at that particular time and surroundings, and to bring back the essence of the originals.

Looking from the distance of time, it can be said that Einstein's yearly scientific period began sometime around 1900, which was marked by the publication of his paper "Conclusions from the Capillarity Phenomena" in the prestigious *Annalen der Physik*.

The end of the nineteenth and the beginning of the twentieth century were a time of great scientific discoveries that were guided by James Maxwell's publication of 1865 *A Dynamical Theory of the Electromagnetic Field* and by the results of the Michelson-Morley experiments.

In 1905, at the age of twenty-seven, Einstein published his four groundbreaking articles: "On a Heuristic Viewpoint Concerning the Production and Transformation of Light," "On the Motion of Small Particles Suspended in a Stationary Liquid, as Required by the Molecular Kinetic Theory of Heat," "On the Electrodynamics of Moving Bodies," and "Does the Inertia of a Body Depend Upon Its Energy Content?" At that time, it was not instantaneous - the Annus Mirabilis (the miracle year) - as some articles are referring to it now. Then the author was an unknown newcomer, and the papers were collecting dust. Almost ten years later, the famed quantum theorist Max Planck brought to light Einstein's work by backing up some of the assertions.

At this point, it is time to bring to light the serious misinterpretations of Einstein's viewpoints on relativity of motion. The first misinterpretation was Einstein's initial consensus with Henri Poincaré's view on relativity.

At the end of the nineteenth century and the beginning of the twentieth century, a French mathematician, theoretical physicist, engineer, and philosopher of science, Jules Henri Poincaré, was the key fatherly figure in the scientific community. His statement of relativity "The all measurements taken at rest or in motion would be indiscernible from each other" was accepted as an axiom.

In reality, in the book *On the Electrodynamics of Moving Bodies*, Einstein has proved otherwise. The article was revolutionary because it demonstrates the abstract nature of time and its applications to the measurements of constant velocity motions. This groundbreaking point of view made it possible to think of relativity in terms of the unified laws by reconciling Maxwell's equations of electrodynamics with Newtonian laws of mechanics.

In this respect, it is important to comprehend Einstein's position on the correlation between constant velocity motions and the nature of electromagnetic media (i.e., light). In his thought experiment on "simultaneity," Einstein clearly demonstrated that the beam light traveling inside of the carriage does it independently from the motion of the system of inertial objects (i.e., the carriage and an inside observer). In other words, he (Einstein) demonstrated that constant velocity motions are distinguishable from one another and from the relatively stationary position as well.

To attribute Einstein with the idea that he was the only one who recognized a "motion of acceleration" is a totally outrageous fiction.

The current interpretation of Einstein's thought experiment "Gertrude's clock" is another example of misinterpretation. At the time of the "birth" of this thought experiment of moving clock and objects, the propagation of any electromagnetic media, such as light, was described as propagation of corpuscular electromagnetic wave. Hence, Einstein envisioned this process as the continuous wave propagation in free space. By trimming this vision of a free space into the "motion of a box with photons moving up/down," the modern interpretation of Einstein's idea of "Gertrude's clock" creates conditions where, at the certain speed of the box, light may miss the mirror at total default of the experiment.

Finally, the biggest problem, as of today, is the inability of the modern scientific community to continue with Einstein's work by providing solid ground for the theories he had developed during his life.

Among those, there are Einstein's questions regarding mathematical apparatus in support of the General Relativity theory, resolution of ether conundrum, etc.

The most important of them all is the one that can resolve the contradiction between Einstein's Special and General theories of relativity.

Is "TIME" an abstract entity that he (Einstein) described during his extensive work on Special Relativity?

Is "TIME" the element of space-time continuum, with all related attributes of a matter, such as stress, etc., which he described in his later works on General Relativity?

It would be a grave mistake to see the first half of the twentieth century as a solid ground of the "carved-in-stone fundamentals" upon which the development of current understandings of science took place. There were, and to some degree still are, unresolved conundrums of ether, nature of gravity, and nature of photon. As a matter of fact, the characteristics of photon were finally defined in 1970. The nature of some kind of a substitute for ether is still not resolved. Moreover, the solution to the nature of gravity that was partially developed utilizing Hermann Minkowski's generalization of rotational invariance from space to space-time, tensors mathematics, and unsettled assumptions of the principles of equivalence is still too tenuous to be accepted as the solution.

We must look at the work of Einstein through the prism of obviousness of those times and appreciate his vision and talent in respect to the level of the entrenched knowledge that was available then.

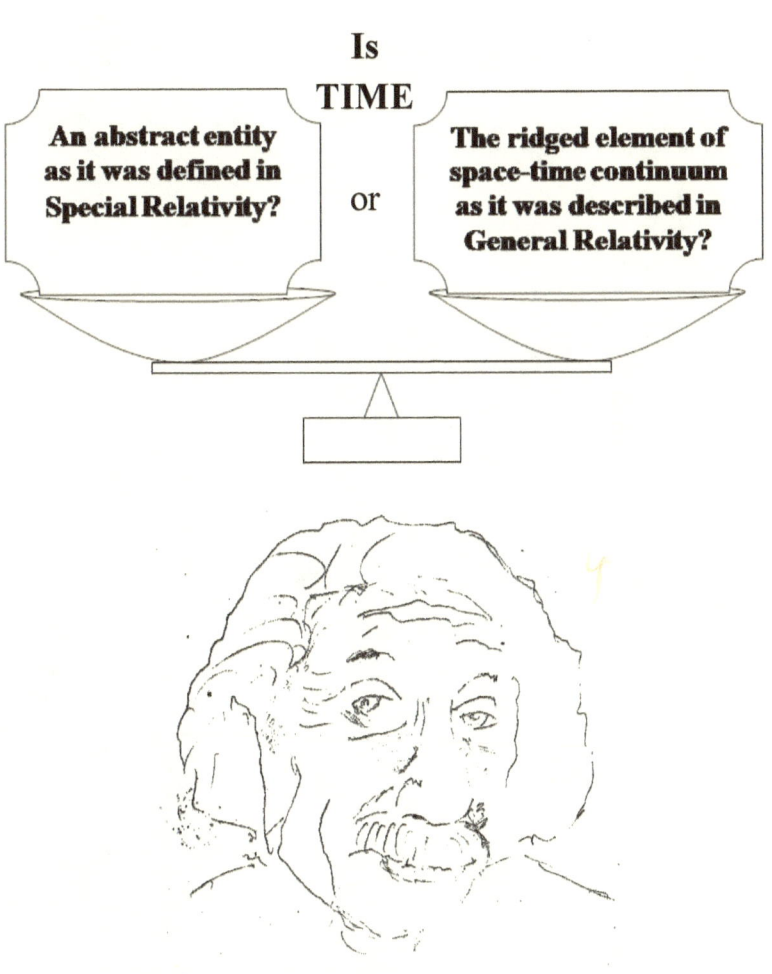

THAT IS THE QUESTION.

?

PART I

AT THE CRADLE OF ELECTRODYNAMICS

On the Wings of Success

FROM JAMES CLARK MAXWELL TO ALBERT EINSTEIN

The Classic Theory of Electrodynamics

From the beginning of civilization and up to the beginning of the nineteenth century, every work in Natural science has described the behavior of inertial bodies. In 1819, Hans Ørsted discovered an existence of the magnetic field around a conductor that carried electric current. Michael Faraday greatly extended his work on electromagnetism, which, by the end of the nineteenth century, was finalized by James Clark Maxwell in his work *Treatise on Electricity and Magnetism*. The first time in human history, James Maxwell has described the nature that governs this massless and noninertial media. All this work on electromagnetism was so different from Newtonian mechanics of inertial bodies that this part of physics became a totally separate and independent entity called - Electrodynamics. James Maxwell not only devised a mathematical apparatus that allowed us to describe the behavior of electrodynamic systems, but he also theoretically established its universal constancy in free space along with the uniqueness of the speed of light.

In 1865, the publication *A Dynamical Theory of the Electromagnetic Field* by James Maxwell could be considered as the date of birth of Classic Electrodynamics. For all practical reasons, Maxwell opened the world of electromagnetism with all its attributes of particle waves, universality, and constancy of the speed of light, along with its independence from any inertial medium.

In 1905, Einstein published *On the Electrodynamics of Moving Bodies*, where he proposed to treat time as a tool of measurement, dismissing the notion of absolute rest. Using Maxwell's electrodynamics and Lorentzian transformation, Einstein has built the bridge between the Newtonian world of physical bodies and Maxwell's world of electromagnetic media, establishing relativity of time.

In other words, Einstein established that any measures, including measures of spatial dimensions and time, are relative. Moreover, Einstein/Lorentz transformations established a correlation between the speed of an inertial frame of reference in space and the measures of spatial dimensions and time made within this or any other such frame.

In 1905, the famous Principles of Relativity were publicized by Einstein in the book *On the Electrodynamics of Moving Bodies*.

With a real stroke of genius, Einstein transformed time from the universally constant entity into an abstract notion or a tool of measurements. By doing so, he established the very base from which the scientific community has built its new identity - "The Quantum Physics" - and the very base that he sacrificed on the quest for his "new ether."

The Classic Theory That Was Contradicted by the Classic Experiment

Maxwell's idea of some medium "ether" (luminiferous aether), through which electromagnetic waves are propagating, became a tripping point for all alternative theories of the nineteenth and twentieth centuries.

The results of the practical experiment by Albert A. Michelson and Edward W. Morley struck right smack into the theoretical assumptions.

This is one of the history lessons when wrong assumptions and misread experimental results misguided scientific community into the search of miracles.

To verify the assumption of luminiferous aether, Michelson and Morley performed the famous experiment that is recognized today simply as "Michelson-Morley" or M-M. The background of the Michelson-Morley experiment was devised based on the general beliefs that electromagnetic wave, just like water wave or sound wave, propagates through its own medium - "luminiferous aether." This ether was assumed to be present everywhere, even in free space, and hence, the Earth passing through it should create the ether wind.

In 1887, Michelson and Morley set up the experiment to detect and measure the influence of the ether wind. In this experiment, a beam of light was directed along the two paths of equal length that was set in right angle to each other (see Fig. 1.2). The apparatus was developed to detect the time difference between the arrival of these beams of light while they are traveling along these two equal paths.

In the explanatory portion of their experiment, M-M provided analogy of their experiment with a boat that travels in a stream of water (see Fig. 1.1). The obvious conclusion from this analogy was that the sums of speeds measured when the boat is moving up or downstream is different from when the boat is moving the same distance across the stream.

Fig. 1.1

Based on the above mechanical model, Michelson has devised the following optical apparatus, which was used in the M-M experiment. The simplified model of which is presented below.

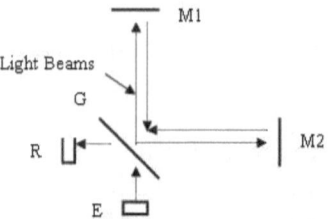

Fig. 1.2

In the device, light emitted from point "E" is split by "G" into two equal beams, which are traveling equal distances between point "G" and mirrors "M1" and "M2." On the way back, the two reflected beams are recombined and fed into the interferometer, "R." Based on the mechanical model, if the "ether wind" does exist, one of the beams will travel slower than the other and the interferometer, "R," will measure this noted difference.

To everyone's surprise, the Michelson-Morley experiment had shown no measurable results.

During the following century, number of attempts were made to reproduce the experiment using most modern technologies, and they had failed to give an explanation of "0" results using everything from Fitzgerald's space contraction to absolute vacuum.

Theoretical Reality of the Practical Experiment

The very base of the problem in the Michelson-Morley (M-M) experiment was laid not in the way of how it was conducted but in the way of how it was conceived and interpreted.

In establishing the theoretical base for the experiment, M-M neglected to provide the proof that:

    a. light is a different matter than ether,
    b. flow of ether directly skimmed the Earth, and
    c. existence of natural observable phenomena that supports assumption (b).

Let us look closely at points a, b, and c:

a. M-M model describes the motion of one entity (a boat) that is moving on the surface of the other and a significantly different entity (water). If light happens to be similar to ether matter, the model should be devised to measure the speed of the process of dissipation of a media disturbance. In this case, the results of the experiment may be interpreted differently. For instance, the speed of the disturbance's dissipation that is identical in all directions indicates consistency and uniformity of the media.

b, c. If ether directly skims the surface of the Earth, we should be able to observe significant amount of natural effects. For instance, in places where the ether wind contacted the Earth, we should expect to identify the areas of excessive pressure. In the opposite end, in place where ether leaves the surface, some kind of ether vacuum should be observed.

If we are to make the proper analogy of the M-M experiment, we may visualize a large ship with a swimming pool filled with outboard waters and an observer who is measuring the speed of dissipation of a water wave that was created when a stone has been dropped into the pool. The results of the measurements should have proved that the speed of the disturbance' dissipation in different directions would vary when the ship is under constant velocity motion versus when the ship is remaining stationary.

Even if we to discard all our objections, we have to admit the following:

The result of Michelson-Morley experiment established only the absence of "ether wind."

Based on the original intentions of the "Search for the ether wind" upon which the experiment was devised, the outcome of the experiment should have resulted in the following statement:

The **Ether wind** was not detected.

The further research in this matter would have taken a totally different approach in solving the dilemma. Instead, the whole scientific community rushed to declare the obsolescence of ether. From this point, ether became heresy of modern physics, and empty space started to curve inward with the unfortunate resolute of the imprisonment of "TIME" behind bars of the space-time continuum. The very person who freed "TIME" from eternal, permanent affixing suddenly sacrificed that freedom for one purpose only—to develop an alternative to the "Ether" model.

THE WORLD OF FUNNY MODELS AND WITTY FANTASIES

First Step toward the "Dark Side"

In September of 1905, Einstein concluded that the law of conservation of mass is a special case of the law of conservation of energy. To prove his theory, Einstein devised the thought experiment that became a part of the physics curriculum worldwide. Below is an example of this thought experiment quoted from "Physics" by Paul A. Tipler:

Einstein arrived at Equation $E = mc^2$ through a simple "thought experiment." He imagined a closed box of mass M and length L. A light flash is emitted from one end of the box. We found that light-carrying energy E also carries momentum $p = E/c$. Initially Einstein's box is at rest with zero momentum. When the light flash leaves one end, it carries momentum E/c. To conserve momentum, the box recoils in the opposite direction (Fig. 2.1).

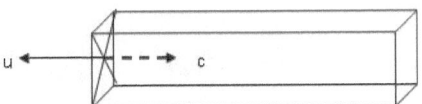

Fig. 2.1

If the box is massive, its recoil velocity, "u," will be small compared with "c" so that momentum conservation may be written as

$$Mu = \frac{E}{c} \quad (2.1)$$

or

$$u = \frac{E}{Mc} \quad (2.2).$$

The light then moves down the box, taking a time

$$dt = \frac{L}{c} \quad (2.3),$$

where again we assume that the box speed, "u," is much less than "c" so that the distance traveled by the light is approximately "L." In this time, the box moves a very small distance, "dx," given by

$$dx = udt = \frac{EL}{Mc^2} \quad (2.4).$$

Then the light flash hits the end of the box, transferring just enough momentum to bring the box to a stop.

But now the box is in a new position. It looks as if its center of mass has moved, yet the box is an isolated system whose center of mass cannot move.

<u>To escape this dilemma, Einstein assumed that light carries not only energy and momentum but **mass** as well</u>!

If "m" is the mass carried by the light, we must have

$$ml = Mdx \quad (2.5)$$

In order for the center of the mass of the system (box + light) will not move. Using our expression for dx and solving for "m" gives

$$m = \frac{Mdx}{L} = \frac{M}{L} \times \frac{EL}{Mc^2} = \frac{E}{c^2} \qquad (2.6)$$

Or

$$E = mc^2 \qquad (2.7),$$

Where "E" is the energy of the light and "m" is its equivalent mass.

Although Einstein's "thought experiment" cited by Paul A. Tipler has number of unsubstantiated assumptions such as

- ➢ "when the light flash hits the end of the box, transferring just enough momentum to bring the box to a stop" and
- ➢ "the box is an isolated system whose center of mass cannot move,"

the most important is the fact that Einstein associated linear momentum, "p," which was assigned to the energy transferred by a stream of light (noninertial medium), with mechanical momentum of the inertial system. It is a well-established fact that the energy of a photon can be described as

$$E = h\nu \qquad (2.8),$$

Where h—universal constant called Plank constant
 ν—frequencies of light.

Somehow, for unknown reasons,
 E (= hν) is associated with $E(=cp)$
 of linear momentum of magnitude "p" (2.9).

A. P. French, in his book *Special Relativity*, describes in this way some background of this assumption:

The only experiments deliberately designed to test the energy-momentum relation have been made not with individual photon but with continuous beam of light, in studies of the radiation-pressure phenomenon. Such experiments involve the incidence of huge numbers of photons (for example, 1 watt of visible light represents a flow of about 3×10^{18} photons/sec) and can be adequately described and analyzed in terms of steady flow of radian energy, without reference to the photonic structure of the radiation.

The radiation-pressure experiments were regarded primarily as verification of Maxwell's theory. However, given a photon picture, they also imply that Eq. (2.9) holds for individual photon.

All radiation-pressure experiments are basically alike. They consist in measuring the force F exerted on a surface by a known flux (measured by the incident power W) of radiant energy. The surface in question is a thin metal vane suspended on a delicate torsion fiber; the energy flux is measured by its heating effect. The experiment can be regarded as a test of the following relation, arising from Eq. (2.9):

$$c = \frac{W(1+q)}{F} \qquad (2.10),$$

Where

"W" is, of course, the rate of arrival of energy, "F" is the rate of change of momentum of the radiation, "q" is reflection coefficient.

It is apparent from the above material that the radiation-pressure experiments are proved beyond any reasonable doubt by only the Maxwell theory of the constancy of the speed of light.

The following two statements are simply overstretched conclusions without serious experimental proof:

- ➢ "The radiation-pressure experiments were regarded primarily as verification of Maxwell's theory. However, given a photon picture, they also imply that Eq. 2.8 holds for individual photon."
- ➢ "The experiment can be regarded as a test of the following relation arising from Eq. 2.9."

More important is the fact that the linear momentum, "p," cannot be equated with the mechanical momentum nor can it be associated with the mass of a body, which is a common character of an inertial system.

An even more obvious blunder was laid in Einstein's assumption that light carried not only energy and momentum but also **mass**, since it is a well-established fact that photons have zero mass and zero charge.

The results of our study of Einstein's "thought experiment" on the "Transformation of Mass in Energy" demonstrate that this assumption of the great physicist must be reevaluated in the view of our existing knowledge. Moreover, in today's advance-technology world where speed of light can be slowed down to a crawl, the idea of the massless light due to its speed has no sense at all.

> **The following is the summary of Einstein's wrong assumptions:**
>
> 1. The very hypothesis that the process by which light is emitted from the "box" creates recoil of the "box" assumes that light has a mass, hence it is inertial media.
> 2. The non-existent relation between "linear momentum" of light and mechanical momentum of the body provides for direct correlation between energy and mass.
> 3. All of the above provide for direct transfer of the energy of motion into mass and in reverse, a mass in motion into energy. Such direct and simple transfer has not being recorded in nature.
> 4. Most important of all is that - we cannot apply Newtonian law of mechanics (intercalation of inertial bodies) to the Maxwellian world of electrodynamics (non-inertial media).
>
> All of the above simply stated that "E" is not "= mc^2"

Therefore, the work of Einstein's *"thought experiment"* on mass and energy must be put on hold till such relation is established.

As it would be apparent from the further discussions in the Book, the idea of Energy as the main building block of Nature is right and the wrong method of describing this phenomenon has yielded unsubstantiated results.

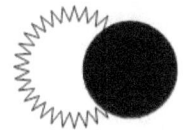

Into the "Dark Side"

Up to this point, we have witnessed a natural progression of scientific theories that are supported by corresponding experiments and vice versa (except the "famous" $E = mc^2$).

From this point, we will travel in the world of imagination, which was constructed only by someone's authoritative mind and with the help of the manipulations of mathematical terms.

This metamorphosis took its roots in the work of the professor of mathematics Hermann Minkowski. In one stroke, Minkowski originated the "U-turn" from the normal progression of science, which was advancing from Newtonian absolute space and time to Einstein's relativistic space-time into the phantasmagoria of rigid net of space-time continuum.

Minkowski's Space-Time

Hermann Minkowski was a professor of mathematics at the University of Göttingen and Einstein's onetime professor of mathematics in Zurich. As the story went, Minkowski had underrated Einstein and was very surprised learning of Einstein's success in developing Special Relativity. Regardless of his initial judgment, Minkowski recognized potentials within Special Relativity for opening of a new dimension - the dimension of time.

In 1908, Minkowski published his work named *Space and Time*. In his prologue to the book, Minkowski wrote the following:

> *Space by itself, and time by itself, are doomed to fade away into mere shadows, and only a kind of union of the two will preserve an independent reality.*

The rationale behind Minkowski's approach was obvious from the very beginning when, in Chapter 1, he chose to base his work on Newtonian mechanics, even though the rest of his rationale came from electrodynamics and Special Relativity. Minkowski wrote the following:

> *The equations of Newton's mechanics exhibit a two-fold invariance. Each of them by itself signifies for the differential equations of mechanics a certain group of transformations. The existence of the first group is looked upon as a fundamental characteristic of space. The second group is preferably treated with disdain, so we . . . of never being able to decide, from physical phenomena, whether space, which is supposed to be stationary, may not be after all in a state of uniform translation. Thus, the two groups side by side lead their lives entirely apart. But it is precisely when they are compounded that the complete group, as a whole, give us to think.*

In contrast to this statement, Special Relativity, contrary to the Newtonian approach, has established relativity of space and time. And doing so, it has transformed space and time from the status of objects into the status of the tools of reasoning. The above statement of Minkowski demonstrated his total disregard of this fact. Apparently, his goal was to reinstate space-time back to the class of objects. He succeeded in his search by using purely mathematical substitutions and unexplainable revelations.

In the same Chapter 1, Minkowski, with ease, made number of sudden jumps, such as from the space-time coordinates used as a measuring tool into the space-time grid of reality and from the application of equations of linear mechanics into the field of electrodynamics of the propagation of electromagnetic waves.

These actions have no explanation in his (Minkowski) work and not even in Sommerfeld's notes.

Below is Minkowski' rationale of such transformation:

> *Let x, y, z be rectangular co-ordinates for space and let t denote time. The object of our perception invariably included place and time in combination. A point of space at a point of time that is a system of values x, y, z, and t I will call a world-point. The multiplicity of all thinkable x, y, z, t systems of values we will christen the world. Let the variations of dx, dy, dz of the space co-ordinates of this substantial point correspond to a time element dt. Then we obtain an image of the everlasting career of the substantial point, a curve in the world, a world-line. The whole universe is seen to resolve itself into similar word-lines, and I would fain anticipate myself by saying that in my opinion physical laws might find their most perfect expression as reciprocal relation between these world-lines.*
>
> *To establish the connection, let us take a positive parameter c, and consider the graphical representation of*

$$c^2 t^2 - x^2 - y^2 - z^2 = 1 \qquad (2.11).$$

Using the rationale of the space-time relativism on which Minkowski based his theory in conjunction with laws of Special Relativity, one should have asked the following question of space and time:

> **Which time and space were Minkowski taking into account?**

Since, as it was established by the "relativity," there are no absolute space and time and because everything is in motion, these are the questions:

- Which time is prevalent, and which space is prevalent?
- What is the space?

It was a well-known fact then, just as it is today, that *space* is an abstract that describes *"that which contains and surrounds all material bodies"* (*Webster's Dictionary*). Yes, we can describe processes in space-time coordinates or describe spatial measurements as well as measurements of time. Fortunately, that alone does not give us rights to speak of space-time as an entity.

Further in the article, using his skills in geometry and without any details and rationale, Minkowski built a brunch of hyperbola corresponding to the expression of:

$$c^2 t^2 - x^2 = 1 \text{ (see Fig. 2.2)} \qquad (2.12)$$

with correlation of OC' = 1 and OA' = 1/c.

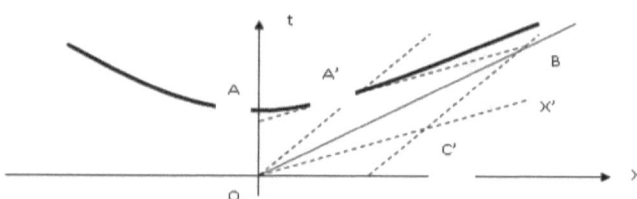

Fig. 2.2

In his work, at the end of Chapter 1, Minkowski made the following statement, which should have stirred up emotions in the scientific community:

> *If we now allow c to increase to infinity, and 1/c therefore to converge towards zero, we see from the figure that the branch of the hyperbola bends more and more towards the axis of x. In view of this it is clear that group G_c in the limit when c = inf., that is*

the group G_{inf}, becomes no other than complete group which is appropriate to Newtonian mechanics.

I will state at once what is the value of c with which we shall finally be dealing. It is velocity of light in empty space.

Here are the questionable assumptions that was used by Minkowski in his work:

- **The speed of light in a vacuum is not a universal constant and can vary from zero to infinity.**
- **Newtonian mechanics equates with Maxwellian electrodynamics.**
- **Newtonian Mechanics exists at infinity of the finite speed of light.**

Unfortunately, the statement of the changeable speed of light in a vacuum has been accepted just as an objectivity of the space-time grid.

Minkowski's choice of mathematical tools of solid mechanics in solving problems of electrodynamics and vice versa not only provided unequivocal support for Einstein's "$E = mc^2$" but also opened the door for even more drastic changes that resulted in the creation of Einstein's General relativity.

> Summarizing all the above, I would like to underline that the application of the space-time approach as a measuring tool has greatly enhanced our ability to record and analyze objective processes. The very fact of the existence of such measuring tool, as an inertial frame of reference, is the result of the application of the space-time coordinates. Unfortunately, Minkowski's theory has a pointless rationale, which transmutes space-time measuring coordinates into the objective space-time grid of actuality.

Note: Due to the unfortunate demise of Minkowski, a more detailed explanation on the work can be found in the "Note" by A. Sommerfeld.

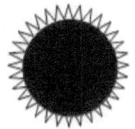

On the Dark Side

The Vital Lesson to Researchers:

"Appearance of Reality is not the Reality of Appearance"

In relation to the light in a Rocket: the strange interpretation of Einstein's visualization of the "behavior" of light inside of fast, upward-moving bodies (e.g., an elevator or a rocket) was clearly explained to me by Prof. Hans Bethe during our conversation about Gertrude's clock. As per Professor Bethe, in the time of visualizing the process, Einstein recognized light as a continuous electromagnetic wave that is significantly different from today's understanding of quantum electrodynamics.

My enlightenment on Einstein's point of view in his "thought experiments" came from my conversation with Prof. Hans Bethe after I have described my objection that at a certain speed of the Light-pulse clock, it will stop working, since light will miss the receiver. That discovery became the base for my work on Velometer that further became the ground for the work on the Technology of Noninertial Navigation.

With the Professor's remarks that were at the beginning of the twentieth Century, the time when Einstein visualized his "thought experiments," the nature of light was understood and described as continuous electromagnetic waves. Hence, the propagation of light was visualized from this point of view.

In the middle of the twentieth Century, the further work on the nature of light has been progressed into the world of massless photons. Unfortunately, Einstein did not come back to reevaluate his vision of Nature that was formed at the beginning of the Century.

That vision became the base for Gertrude's clock and the vision of a "bending" light inside of a rocket, which became the stepping stone to General Relativity.

First Assumption of General Relativity

The cornerstone of Einstein's General Theory of Relativity is the Principle of equivalence, which equated the gravitational field with uniform acceleration. If such principle is true and the gravitational force is the function of an object's uniform acceleration, all physical bodies in the universe should have reached the limit of the speed of light, "c," a long, long time ago. Apparently, it is not so. Moreover, when Einstein described influence of acceleration on a beam of light in his "thought experiment," he used only the appearance of the reality in the momentous snaps of the overall continuous process.

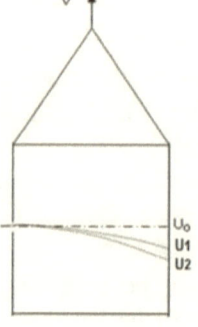   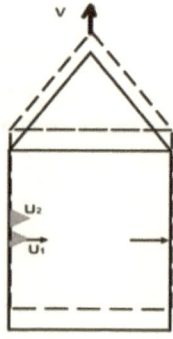

Fig. 2.3   Fig. 2.3a

According to Einstein's "thought experiment," Fig. 2.3 depicts a spaceship under motion, where a beam of light from an illuminator on the left wall illuminates a spot on the opposite wall.

First line "u_0" depicts the position of the beam of light at the ship's velocity – U=0. After that, the spaceship under the uniform acceleration reaches the speed "U1" and the beam of light trajectory depicted by "u_1." Since Einstein equated the uniform continuous acceleration with gravity, we have to assume that the spaceship will continue its acceleration. In this case, the beam of light will continue "to bend," and its trajectory will be depicted by "u_2," when the spaceship reaches velocity "U2" and so on.

Hence, as a result, the light path will continuously bend under uniform acceleration, which should indicate that gravity will continue to increase under influence of this uniform acceleration. Therefore, gravity should be continuously a variable number unless acceleration is ceased.

Replacing an Appearance of reality by the Reality of the process, we will analyze the same Fig. 2.3 utilizing known and well-described laws of propagation of light within an inertial frame in motion.

According to Special Relativity, a beam of light will travel within the body in motion in a straight line and is independent from the motion itself.

In this respect, it is important to consider that based on the laws of electrodynamics and Einstein's Special Relativity, the propagation of light within the body in motion can be described as follows:

- ➤ **Not only acceleration but the velocity of the spaceship influences the displacement of the point of impact of the beam of light on the opposite wall of the spaceship.**
- ➤ **A light trajectory in vacuum will not be curved under motion, and it will continue to be a straight line. There**

are no known forces that will significantly alter the path of light in a vacuum. Even gravitational fields of large celestial bodies may alter a light path, an element of the trajectory, at infinitesimally insignificant amounts.

Hence, the trajectory of the light inside the spaceship will not be a curve but rather a straight line.

Einstein's General Relativity

In 1911, Einstein derived the "Equivalence Principle," which later became the foundation for General Relativity.

Why did Einstein decide to abandon the freedom of the independent space-time of Special Relativity for the rigidity of the space-time grid?

We can only guess that the negative result of the Michelson-Morley experiment, along with the authoritative simplicity of Minkowski's four-dimensional world, played some role in Einstein's decision. Once he (Einstein) made this decision, he realized that Lorentzian transformations could not work under this assumption; nevertheless, he discarded this *"little discrepancy"* in favor of his quest.

Frame of Reference

Over the last century, the notion of a frame of reference became somehow distorted, so in this chapter, we will clarify that.

In *Theoretical Physics*, Georg Joos addressed this question as follows:

> 1. *Frame of reference—the tacitly assumed existence of a frame of reference (coordinate system) in which we could specify the position of an object under study from instant to instant by giving certain number and, further, a means of measuring time (a clock), which marked off definite*

equal intervals of time at which the position could be recorded.
2. *Inertial frame of reference*—a frame in which Newton's laws are valid.

From the above, it is clear that we can have only an inertial frame of reference, if we are to assume the existence of a noninertial frame with space-time coordinates that is moving along the x-axis with the speed $v \to c$.

Based on Fitzgerald's space contraction, the following will take place:

$$x_1 = x_0 \sqrt{1 - \frac{v^2}{c^2}} \quad x \to 0 \qquad (2.13),$$

And

$$t_1 = \frac{t_0}{\sqrt{1 - \frac{v^2}{c^2}}} \quad t \to \infty \qquad (2.14).$$

A noninertial frame cannot exist because the measurements cannot be made along the axis where space is shrinking to zero and time is stretching to infinity.

M. S. Longair, in his book *Theoretical Concepts in Physics*, explained the reasons for Einstein's abandoning of Euclidean geometry as follows:

> *The simple argument indicates that, according to the principle of equivalence, the geometry along which light rays propagates in an accelerated laboratory is, in general, not flat but curved space, the spatial curvature depending upon the local value of the*

*gravitational acceleration **g**. Notice that this argument only makes it plausible that spatial section through space-time should be curved. In the full theory, however, the whole of the metric will refer to a curved geometry, i.e. four-dimensional space-time should be described by curved four-dimensional geometry.*

At this point, a choice has to be made to find the mathematical apparatus that will fit this new physical model. Einstein, with the help of his friend mathematician M. Grossmann, found the answer in Riemannian geometry, Minkowski's space-time, and metric tensors.

The result of Einstein's vision of gravity as a curved space-time continuum became known as Einstein's field equation:

$$E_{\mu\nu} = \frac{8\pi G T_{\mu\nu}}{c^4} \qquad (2.15),$$

Where:

E_{mv}—Einstein curvature tensor (derived from "Riemann curvature tensor"),
T_{mv}—The energy momentum tensor (source of gravity),
G—Newtonian gravitational constant,
C—Speed of light,
m and v—Possible coordinates of space and time.

> As it is well researched and accepted, Gravity is always treated as a gradient field. On the other hand, "Tensors" are mathematical apparatus that are devised to analyze multidimensional arrays, or an object called the "tensor field," often referred to simply as "tensors." Tensors are used for calculation of gradients in vector fields, etc. Hence, it has to be proven that Gravity is an element of vector field; otherwise, it cannot be derived from or described by tensors metrics.

The Equivalence Principle

In the article titled "On the Influence of Gravitation on the Propagation of Light" published in *Annalen der Physik*, 35 in 1911 A. Einstein used the following model of two systems where a stationary system of coordinate K is under influence of a gravitational force that run in a negative direction along the Z axis. At that, he (Einstein) compared this system with the "similar" frame of coordinate K' system that is moving with acceleration in the gravity-free space in the positive direction along the Z' axis.

At this point, Einstein made the decision to sacrifice his crown jewel of "Special Relativity" and declared that he will analyze the system from the point of view of Newtonian mechanics!

Moreover, he (Einstein) concluded that both systems are identical.

At first glance, it seems that these two models (K and K') are identical, but are they?

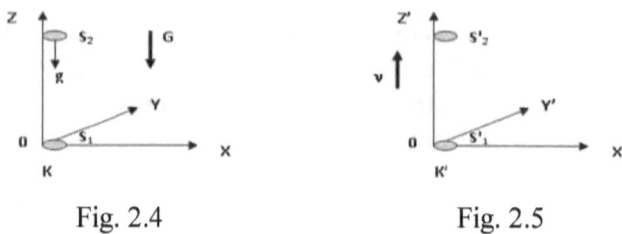

Fig. 2.4　　　　　　　　Fig. 2.5

Since Einstein stipulated that the systems of coordinates K and K' are obeying the Laws of Newtonian mechanics, at that, they are continuing to be **Inertial frames of reference**.

Problem 1

The inertial frame of references K is stationary, but K' is in motion and, even under the Laws of Newtonian physics, forces and momentum that acted on separate components S_1 and S_2 versus S'_1 and S'_2 in the stationary frame are different than in the frame in motion.

Problem 2

If systems S'_1 and S'_2 are parts of the inertial frame K', as the corresponding systems in the frame K, the Laws of Newtonian mechanics require that the systems S'_2 will have the same acceleration as the system S'_1. Otherwise, the system S'_2 is independent of the frame K' and the Fig. 2.4 and Fig. 2.5 are not identical.

- If the system S'_2 is a part of K', it must move with the same velocity, and we observe no changes in the distance between the two material systems.
- If the system S'_2 is not a part of the frame K', it must be suspended in free space with velocity $u = 0$, and the two frames are not identical.

Problem 3

As per Einstein's model, the inertial frame K is stationary and "material" systems S_1 and S_2 are under the force, F_k, of gravity, G.

Even though Einstein did not specify the source of the gravity, we can assume that under a described condition, the gravity was created by a material system with mass m_k significantly larger than the masses of the material systems S_1 and/or S_2.

Hence, gravitational force created by the mass, m_k, is:

$$F_k = -\frac{G(m_{s2} m_k)}{r^3} r \qquad (2.16),$$

Where:

F_k—force,
G—gravity,
m_{s2}—mass of the material system S_2,
m_k—mass of the unnamed gravitational system,
r—distance between S_1 and S_2 (since S_1 is stationary under the gravity).

Under that force, the system S_2 is moving within the frame K with acceleration, g, in the direction of $-z$.

On the other hand, the inertial frame K' by itself is moving under the acceleration, n, in the direction of $+z$.

Consequently, through the laws of Newtonian mechanics, we can determine the following:

The force $F_{k'}$, which accelerated frame K' and the system S'_2, is:

$$F_{k'} = m_{s'2} v \qquad (2.17)$$

And

$$F_k >> F_{\bar{k}} \qquad (2.18).$$

Problem 4

Although Einstein simplified his thought experiment by restricting it to the purely mechanical processes, in the end, the experiment should adhere to the principles of relativistic mechanics. The laws of relativistic mechanics make the difference between the frames even more significant.

Relativistic masses in the two frames are totally different, for instance, the mass $m_{s'2}$ suspended in free space system S'_2 is the idealistic mass m_0 and is the smallest mass of all. The mass $m_{s'1}$ of the system S'_1, which is only under influence of the force of acceleration, is the second lowest. The mass m_{s1}, which is relatively stationary under the force of gravity system S_1, is the next to the last, and mass m_{s2} of the system S_2, which is accelerated under force of gravity, will be the highest:

$$m_{s2} > m_{s1} > m_{s'2} > m_{s'1} \qquad (2.19).$$

All these relativistic masses corresponding energy levels, etc. . . . will be different from one another. As a result:

- in the frame K, the system S_2 under the influence of gravity is moving toward the system S_1, relativistic mass and energy, of which is lower than its own
- in the frame K', the system S'_2 is stationary and the system S'_1, under the influence of a physical force, is moving toward the system of lower than its own relativistic mass and energy.

Conclusion: it is obvious from the above that even though numerical equivalence of the acceleration, created by the force of gravity in the frame K and the acceleration, created by the force applied to the frame K', the physical processes within the frames are different.

> **In simple terms, inertial frames K and K' are not equivalent. There is no commonality between these frames, except that the distance r between the corresponding material systems is decreasing with equal rates under the influence of gravity and acceleration, respectively.**

The inertial frames of references K and K' can be equal only if both of these frames, including all components, are in motion. At that K, including S_1 and S_2, under force of gravity "G", is moving downward with acceleration g and K', including S'_1 and S'_2, is under influence of some unknown force moving upward with acceleration "a". Evidently, there is no magic in that.

> **In his (Einstein) strive to solve the problem of gravity through General Relativity, under the existing understanding of Nature, Einstein has neglected the practical rules of a researcher by taking an appearance for the reality and forsaking his own work on Special Relativity.**

As a result of the above, we have to choose between the two distinctly opposing points of view - one that was established by the Principle of equivalency and the other that was established by Relativistic physics.

There is an additional observation that we have to take into account when we are analyzing tools used in General Relativity. The mathematical apparatus, which Einstein chooses, as a tool for manipulation of data in General Relativity, does not belong there and it can't be applied as such. On one hand, we have Space-Time Continuum, which based on Minkowski Space and Time theory, is a scalar entity of a gradient field. On the other hand, we know that tensors are mathematical tools that were developed to describe the spatial displacement of vectors.

Results of the analysis of the "Equivalence Principle," as well as correlation with "Special Relativity", as described above, clearly demonstrate that the "General Relativity" was built based on unfound assumptions.

PART II

AT THE SOURCE OF EVERYTHING

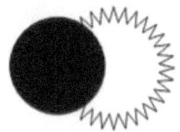

From the Darkness into the Light

Is it possible that in Nature, some things do exist for the sake of existence except some species of humans?

Step by Step and Block by Block

The first step into the "Light" begins with looking back on the point of confusion that was created by the results of the Michelson-Morley experiment and by asking the following questions:

a. **Why is there no ether wind?**
b. **Do we have another plausible explanation for zero wind results of the M-M experiment?**

As it was outlined in the subchapter **"Theoretical Reality of the Practical Experiment"** of this book, in establishing the theoretical base for the experiment, Michelson neglected to provide the proof that light is a different matter than ether.

> **On the practical side, it would be wise to review the well-accepted notion of ether.**

In this respect, there are still unresolved questions:

- Is ether some kind of luminiferous inertial media that fills all or some areas of space?
- Is it possible that some things do exist in nature for the sake of existence?

Our current knowledge of Nature would more likely provide the negative response to these questions.

In the next step, we need to find plausible resolution to what was sited in our first-step conundrum. In this case, I chose to look into the most common manifestations of "Energy" in Nature.

We know that

- ❖ all processes of the transfer of energy, including thermodynamics, are related to the field of electrodynamics;
- ❖ an object (how we describe any matter that possesses inertia) contains significant amount of energy, and it is

very possible that all matters are some kind of forms of energy;
- ❖ matter is always in a state of constant motion that requires some energy; and
- ❖ an object is always in continuous active state where amounts of energy constantly flow in/out of it.

In other words, we can stated that:

> **The process of exchange and containment of energy is essential to the very existence of matter.**

The above statement is the key for understanding the innovative approach I have described in the further chapters, so let's elaborate further on this subject.

Here are the well-established facts of Nature:

- ➢ Energy is invisible and a directly undetectable "substance."
- ➢ We can change matter's internal structure or even matter itself by providing energy to it.
- ➢ Matter can exchange energies, and by doing so, it increases or decreases its own energy level.
- ➢ Energy can be transferred through different kinds of matter and over significant distances in a vacuum.
- ➢ Energy cannot be created nor destroyed; it can only be transformed.
- ➢ Energy is regulated in nature by the laws of balance.
- ➢ Energy cannot disappear or reappear momentarily.
- ➢ Energy is a part of matter regardless of its level of complexity.

From now on, we will describe Energy in terms of abstract, noninert media that can influence matter in all its Levels of complexity.

The new term **"Level of complexity"** will define the level of physical structure of objects. The simplest form of an object that is called a Gluon (as far as we know today) should have complexity no. 1, Quarks no. 2, Neutrons and Protons no. 3, and so on. The level of complexity defines not only internal layers of objects but also the complexity of internal energy exchanges within, as well as the objects' reaction to the outside world.

Taking all the above in consideration, we can say that we have established the initial building blocks of our new model, and it is time to start building our new model block by block, which I have named **"E-model."**

E-Model

The Rule of Exchange

The new model is based on the rule of exchange of energy, which is:

> **For matter to exist, it needs to continuously exchange Energy**

The very existence of matter is the sole function of the energy exchange within a matter, in-between matters, and between matter and the surroundings. Every element of matter, regardless of its size and structure, requires energy. The processes of energy absorption, retention, and exchange are the functions of matter. The amount of such energies is corresponding to the mass and the level of complexity of such matter.

We can imagine a stream of energy that is directed from a surrounding area toward every singular space on the surface of an object (body) and all the way inside to provide energy and to support energy exchange on every level of its complexity from an elementary paticle and on.

Consequently, it is not hard to imagine that the force of gravity inside, on the surface, and in the outward space of an inertial body, created by the stream of energy that is absorbed by the matter of the body.

Moreover, the force that attracts inertial bodies to one another resulted from the Energy Field (E-Field) distortions, created by the flow of energy into each object.

Our new E-model of gravity and gravitational force is developed in accordance with Newtonian law, which states that the gravitational force is proportional to the mass of the body and dissipation of which is proportional to the spherical distance from the body:

$$F = G\, m_1\, m_2 / d^2 \qquad (10.1),$$

Where:

F—gravitational force,
G—gravity,
m_1—mass of a substance,
m_2—mass of a substance,
d^2—distance.

It is easy to understand why gravitational force varies within and about the surface of an inertial body since in our model, gravitational force on the surface and outward space is the function of the body's internal configuration.

The First Positive Conclusion

Our E-model provides clear explanation to zero results in Michelson-Morley experiment.

If a stream of noninert media of semiuniform concentration is coming toward the Earth, the measurements of propagation of noninertial media, such as light, in any direction within this stream should yield identical results.

In other words, it should be obvious that under such conditions, measurements or even existence of the ether wind is not possible. It is perceptible that the Earth, as a large celestial body, acquires proportionally large amount of energy that streams toward its center. This stream of energy, the force of which we call gravity, is somehow reconfigured by the Earth's motion in space. The Earth, while moving through space, continuously pulls that stream of energy. As a result, every measurement that is related to the propagation of noninert media within this stream of unrelated energy will yield an identical result in any direction within that stream.

Establishing E-Model

At present, we have a significant amount of theoretical and practical material that supports the new model, and the details of which will be described in the following subchapters.

First, let us fully describe the new **E-model**, where "E" designates Energy.

"E-model" is characterized by the following conditions:

- Prime energy exists in the universe "E_0."
- Matters are the products of this energy that exists within.
- There is the continuous energy exchange process in the energy/matter system, which are characterized by the following:

    a. Energy absorption
    b. Energy transformation
    c. Energy retention
    d. Energy expulsion

- Balance of energies is eventuality, whereas imbalance is a momentary snapshot of matter.
- Motion of matter in space is the result of the energy exchange process.
- Forces of attraction and expulsion between matters are the result of the energy exchange process.

Energy, E-Field, and Energy Exchange Process

The prime Energy E_0 is the field (E_0-field), in which our universe exists, and it is the basic building material in the universe. The E_0-field is not the hypothetical neutral substance called ether and it is not a mysterious dark matter, that fills up all empty spaces. Rather, it is the foundation of the existence. Since everything in the universe is derived from the E_0-field, it is not possible to detect it. The E_0-field fills some space in the universe and what we call a vacuum, is only described as a state where only some matters are absent to some degree. The E_0-field is absorbed by every element in the universe, and the flow of the E_0-field into a body creates a force, which is proportional to the mass and level of complexity of the body. This force is identifiable, and in the macro systems, it's called - **Gravity.**

Energy Exchange

As our theory describes, the universe consists of inhabited and empty spaces, where the empty space is free from the E_0-field, that may be defined as an "absolute vacuum."

Matter does not exist in an absolute vacuum. The E_0-field occupies all inhabited spaces in the universe.

The unevenness and nonuniformity of the E_0-field resulted from the existence of matter. On the other hand, the nonuniformity is responsible for the existence of matter. Due to the nature of the E_0-field, under some conditions, a simple form of matter can be created from it. After a simple matter is formed, it requires a constant flow of energy from the E_0-field (energy exchange). By doing so, it distorts (influences) the E_0-field in its proximity.

Based on the above, a body will have the ideal exchange conditions when it is stationary in relation to the E_0-field. In this case, the speed of the energy exchange will be a function of the outside energy density, size, and complexity of the body.

Having knowledge of the behavior of an object inside of an electromagnetic field, we can assume that the process of interaction between the E_0-field and the inner object's energies will create a force, which will provide movement to the object. It will work similarly to an eddy current, created inside of a piece of metal placed within a magnetic field. At the same time, the process of the energy exchange will work as a brake force that prevents the object from constant acceleration. The summary of these two opposite directed forces will establish the speed of motion of an object within the E_0-field.

Based on what is described above, we can conclude that as soon as a body (an object within the E_0-field) has formed, the motion becomes a necessary part of its existence. It would require an outside force to prevent an object from the established movement.

At the same time, this movement is slowing down the process of the energy exchange, hence depriving the object from the required amount of energy. Long time ago, Max Plank provided us with the equation that can be used to describe the rate of Energy Exchange process:

$$p = \frac{mu}{\sqrt{1 - \frac{u^2}{c^2}}} \qquad (10.2),$$

Where

>p—rate of exchange,
>m—mass of a substance,
>u—speed of a substance in space,
>c—speed of light in vacuum.

All that leads to the assumption that:

The speed of the energy exchange process proceeds with the speed of light.

Knowing that the speed of light is the speed with which the transfer of electromagnetic energy is taking place in Nature, we can safely assume that "c" will represent the speed of transfer of any kind of energy in a vacuum as well.

At this point, it would be easy to explain why objects, which are moving with different speeds in relation to the speed of light, have different measures of time.

For instance, it would explain why in the *mu-meson* experiment, the measurements had showed longer life.

Based on the above equation where $\frac{u^2}{c^2}$ (the correlation between a speed of a substance in space to the speed of light) is used rather than $\frac{u_1^2}{u_2^2}$ (the correlation between speeds of different bodies), we can conclude that contrary to some derisory interpretations of the Special Theory of Relativity, the Energy Exchange Theory indicates the following:

> **The Energy Exchange process, as well as all other processes and measurements made inside of matters, is influenced by the speed of propagation of such matter in the E_0-field.**

Here are some additional conclusions from the above statement:

1. Constant depletion of energy through the required Energy Exchange process will lead to the end of the life cycle of a matter. Therefore, we can state that on one hand, movement is the necessity of life, but on the other hand, the very same movement is the cause of "end of life."
2. If $p = 0$ (where p is the rate of exchange), life is forever! To accomplish it mathematically, we have to have $u = 0$

(where u is the speed of an object) or m = 0 (where m is the body mass) or both = 0.
3. Speed of the energy disturbance is the top speed in the universe accomplishable only by the massless particles. Movement of an inertial matter with this speed is equal to the disintegration of such matter.
4. Speed of matter in the E_0-field is a function of the mass of the matter and the density of the E_0-field in that particular area that surrounds the given matter.

The Energy Exchange Process is responsible for the universal motion of all objects in space. It is impossible to find a stationary object in the universe.
<u>**Motion is the attribute of any object and in any form from a particle to a galaxy.**</u>

Lorentz Transformations, Plank Equation, Special Relativity, and Their Relevance to the Energy Exchange Theory

More than five centuries ago, Galileo recognized that identical measurements produced in different systems in motion may yield different results. As a result of this discovery, Isaac Newton provided us with the mathematical apparatus (we called Galilean transformation) that negates this difference:

$$r = r' + ut \qquad (10.3),$$

Where measurements that were taking in the inertial frame of references "r" are correlated with the measurements that were made from the inertial frame of references "r" through addition of the product of velocity "u" and time "t," with which inertial frames are moving in relation to each other.

Maxwell's discovery of the constancy of the speed of light "c" brought necessary adjustments to Galilean transformations since propagation of light in every frame should be identical, such as

$$x^2 + y^2 + z^2 - c^2 t^2 = x'^2 + y'^2 + z'^2 - c\, t'^2 = 0 \qquad (10.4).$$

The solution to the problem was found in the coefficient:

$$k' = \frac{1}{\sqrt{1 - \frac{x^2}{c^2}}} \qquad (10.5).$$

Both the 10.2 and 10.5 equations describe the relation between the speed of matter and the speed of light and how the speed of light influences some measurements that are taking place within inertial frames of references.

This finding, along with Maxwell's constancy of the speed of light in a vacuum, became the cornerstone for Einstein's Special Relativity theory. The theory clearly demonstrates the significance of the universal constancy of the speed of light in a vacuum and that the principles of absolute time and space are just an abstraction, untenable and artificial.

It took almost a decade before Einstein's Special Relativity was accepted as the statement of the major law of nature. The reason for this delay was the well-accepted "Principle of Relativity" coined by the French mathematician J. H. Poincaré. Back in 1904, he published *Principles of Relativity*, in which he stated that:

> the laws of physical phenomena must be the same for a fixed observer as for an observer who has a uniform motion of translation relative to him so that we have not and cannot possibly have any means of discovering whether we are being carded along in such motion.

Although the resolution between Newtonian Mechanics and electrodynamics was established more than a century ago, the difference between Poincaré's "Principles of Relativity" and Einstein's "Special Relativity" continues to haunt us even today, even though Einstein, in some of his "thought experiments," used Poincaré's principles to promote relativity of time itself.

Someone may ask the question:

Are these two principles different, and where is the contradiction?

The answer is yes.

Both of these physical statements are significantly different.

- Poincaré's Principles establish absolute equality of inertial frames under uniform motion of translation and their equality with stationary frame, as well as our inability to establish such motion from within.
- Einstein's Special relativity, along with Lorentzian transformations, establishes ways to differentiate between the inertial frames under uniform motion of translation and provide the methods, which will allow performing such a task as establishing a uniform motion of translation of any object from within.

Taking above in consideration, let us establish the ways by which the uniform motion of translation can be detected in relative terms.

We have to emphasize the relative term of the suggested method since absolute rest does not exist. Therefore, all measurements can be done in relative terms only and all frames of reference will have a final unidentifiable direction in free space.

Are inertial frames equal or equivalent?

The answer is yes for both.

The following three principles will help to understand this relation:

> **Principle of Equality**
> Based on Special Relativity's First Postulate, laws of nature apply equally to all matter and within all and any inertial frame(s).
>
> **Principle of Equivalency**
> Inertial frames are not Equal, but they are Equivalent, relative to each other, through Lorentzian Transformations.
>
> **Principle of Correlation**
> Based on the constancy of the speed of light and its independence from any inertial frame, Lorentzian Transformations provide for equivalency and constancy of measurements done in any inertial frame.

It is interesting that these three principles will not only realign and streamline the ways we operate but also help solve some long-standing misconceptions, like the conundrum of the Twin Paradox.

The Twin Paradox

On one hand, time dilation, as it was established by the Lorentzian transformations, dictates that an astronaut who travels through space must be younger upon his return than his twin brother who remained Earthbound. On the other hand, Poincaré's "Principles of Relativity" dictate that since all frames are equal, the Earthbound twin brother is moving away from the spaceship just as the astronaut on the spaceship is moving away from its twin. Hence, the two frames being entirely equal, the Earthbound brother would experience the same influence of time as his space-exploring brother.

Application of our established three principles clears up the contradiction, since all inertial frames are equivalent through the Lorentzian Transformations and

$$t = \frac{t_0}{\sqrt{1-\frac{u^2}{c^2}}} \qquad (10.6).$$

Lorentzian correlator establishes that time dilation is only a function of an object's speed in relation to the speed of light. Using the above, we can derive only one possible conclusion in the "Twin Paradox":

The only possible explanation about the relation between the speed of light and the speed of an object is the physical dependence of the object on correlated processes, which are occurring with the speed of light.

This conclusion unequivocally points into the direction of the Energy Exchange theory.

> **The correlation of the brothers' spatial speed with the speed of light will determine the age of the astronaut as well as his/her earthbound twin.**

Rules That Guide the Processes of the "Energy Exchange" Model

The issues raised in Part 1 of this book regarding Einstein's principles of Special and General Relativity clearly illustrate the need for some unified definition of a noninert substance we call Energy.

Moreover, using the existing level of knowledge, we need to establish fundamentals of its existence and rules of its interactions.

Repeatedly we are hearing that someone or something ran out of energy, acquired some energy, or needed more energy, etc.

We even describe energy in terms of capacity for work as a result of the motion of mass in the system or the configuration of mass or charges in the system or the presence of photons in the system (radiant energy). Energy was represented as a scalar quantity with dimensions:

$$(\text{mass} \times \text{length}^2) / \text{time}^2 \qquad (10.7).$$

It is obvious that the description and the above equation were developed quite some time ago; as a result, they provided us with crude approximation of the real energy transfer processes. The further work on principles of thermodynamics, electrodynamics, and nuclear physics moved our understanding deeply inside of matters. The present state of knowledge attributes the Energy Transfer processes to the fields of Classic and Quantum Electrodynamics. This point of view became the platform for development of new and more modern understanding of the Energy Exchange processes.

At the beginning of the twentieth century, there were two scientific directions being established in the quest to develop a unified theory that will describe the very keys of Nature's function, some kind of a theory of everything.

One direction, we may call it "Nuclear," was headed by Niels Bohr, Max Born, Paul Dirac, Erwin Schrödinger, and others whose work was set to look inside a nucleus.

The other direction, we may call it "Gravitational," was headed by Einstein, with the goal to look outward into the universe to devise the mechanism of gravity. In the end, these two directions should have met/each other at the adjoining borders. Unfortunately, each team chose their own ways and methods for the journey that have yielded irreconcilable results.

Extending Quantum Tools into Outward Space in Our Search for Gravity

Since the Transform Energy process has provided plausible solutions in the practical world of quantum physics, therefore, we may take a chance to extend their application further into the global scale of gravity.

Working on the electron theories, Schrödinger and Dirac have established a foundation of quantum mechanics that was supported by experimental research.

We will concentrate on some aspects of Dirac's work since it will help to formulate our results.

Below is the Lorentz-Dirac equation:

$$m\dot{v}_\mu - 2/3\, e^2(\ddot{v}_\mu + \dot{v}_\nu \dot{v}_\nu \dot{v}_\mu) = ev_\nu F_{inc}^{\mu\nu} \qquad (10.8).$$

Dirac discovered that:

> *it would appear that we have a contradiction with elementary ideas of causality. The electron seems to know about the pulse before it arrives and to get up acceleration (as the equation of motion allows it to do) just sufficient to balance the effect of the pulse when it does arrive.*

Working on the electron theories, Dirac finally concluded by saying,

> *Let us imagine the aether (ether) to be in a state for which all values for the velocity of any bit of the aether, less than the velocity of light, are equal probable. This state of the aether, combined with the absence of ordinary matter, may well represent the physical condition which physicists call a perfect vacuum. In*

this way, the existence of aether can be brought into complete harmony with the principle of relativity.

In 1962, Dirac proposed a new electron theory where he considered an electron as a bubble of electromagnetic field, like a small sphere with the charged surface.

Based on our present state of knowledge, we have to conclude that what is sufficient for the microworld should be sufficient for the macroworld. The point is how to define the process of energy transformation that, in one stroke, will provide the clear model of gravity, the region of gravitational and electrodynamic fields.

The Theory of Dual Transformation

Explanation of the Energy Transformation Process

Contrary to existing theories, our model describes the process of energy absorption by matters as the multistage process that consists of at least two stages.

The First stage is where the E_0-field with continuous spectra is absorbed by matter. That action results in the First Transformation of the E_0-field. The Second stage is where the transformed energy exits matter, which results in the Second Transformation, hence becoming the well-known Electromagnetic Field.

The First Transformation

We have to assume that the E_0-field is everywhere. It has no boundaries and is what we may call the infinite entity with unbounded domain and continuous spectra. As soon as the E_0-field enters a boundary of a substance, it produces some sets of discrete spectra:

$$f(s) = L\{F(t)\}, \quad \text{and} \quad F(t) = L^{-1}\{f(s)\}, \qquad (10.9).$$

In this case, Laplace transformation offered means for treating the problems where the values of the functions were used to form the product solutions of the discrete spectra for bounded domains, but it had a continuous spectrum for an infinite or unbounded domain.

Thus, in bounded domains, we were able to construct solutions in terms of sums over the spectra of these values:

$$f(s) = \int_0^\infty e^{-st} F(t) dt \qquad (10.10).$$

Straight Laplace transformation (10.10) in our case, where

$$\varepsilon = F(E_0) dt \qquad (10.11),$$

reflects the process where the prime "homogeneous" energy of E_0-field reaches a homogeneous boundary and it is transformed into the first level, e-Energy. The amount of energy, "s," being transformed is proportional to the mass, "m," of the substance in the power of its level of complexity, "b":

$$s \cong m^b \qquad (10.12).$$

Substituting 10.12 in 10.10, we can write the first transformation as

$$f(s) = \int_0^\infty e^{-m^b t} F(t) \, dt \qquad (10.13).$$

> **It has to be noted that the First Transformation introduces the function of natural cycles, which we are measuring in terms of "TIME" in all processes of Nature.**

Second Transformation

After the homogeneous energy of the E_0-field was transformed into e-Energy at the point of entering a substance, it becomes an inhomogeneous, time-related entity that continues to be massless.

When e-Energy is leaving the substance, it has to be described as an inhomogeneous energy that is going through the same homogenous boundary.

In this case, Fourier transformation offers means for treating problems with transformed e-Energy:

$$e(t) = 1/2\pi \int_{-\infty}^{+\infty} e^{ikx} \varepsilon(x) dx \qquad (10.14).$$

When transformed energy is released, it produces disturbance with an array of different frequencies because of its inhomogeneous structure.

> **The second transformation introduces an Energy that is known to us as the electromagnetic or e/m–Energy with its spectrum of frequencies.**

There is, more likely, that a matter holds two types of transformed energy within. And when energy is released, we can detect only the second type of transformed energy (a.k.a. electromagnetic disturbance).

We have to admit that the First and Second transformations resulted in providing us with the massless fields of energy. There has to be some other processes that create matter from the massless field.

We can presume that these possible processes can be described through reverse transformations.

Therefore, the reverse transformation in our case is:

$$e = \Phi(\varepsilon)d\eta \qquad (10.15).$$

The above reflects the process of the reverse transformation, by which the contained in the body transformed energy is released into the open space.

It is intriguing to notice that the suggested process creates a disturbance of dual-converted energy, which we may call an electron with a particular mass and a tight connection to the frequency spectrum.

Therefore, we can identify photon as basic quanta of disturbance of transformed Energy and an electron as a basic quantum of disturbance of the dual-converted energy. Since the transformation proceeds in a complex plane, the basic quantum of dual-converted energy has to acquire the complex form such as

$$a + ib \qquad (10.16),$$

Where

$$i = \sqrt{-1} \qquad (10.17).$$

The result of 10.17 provides a mathematical representation of antiparticles.

Gravity as the Energy Absorption Process

The Gauss's Theorem—the surface integral of the vector, "n," taken over a closed surface is equal to the volume integral of the divergence of the "n" taken through the enclosed volume—opens doors for our understanding of the gravitational process of Energy absorption by any object, including celestial bodies:

$$\oint \upsilon dS = \int div \upsilon d\upsilon \qquad (10.18).$$

Is a quantum of energy a particle?

The "Hunt for Particles" that began back in the twentieth century became a predominant force of theoretical and applied science that drives all existing scientific communities of the twenty-first century. Smashing and splashing matters for the sake of searching for idealistic "goddamn particles" became a norm.

There are number of questions that have to be answered before billions more will be wasted in support of this idealism:

- What is a particle, and what is a quantum?
- Is a quantum of energy a particle?
- Can we continue to describe "packets of energy," like a photon, as a quantum, or is it a particle now?
- Do quantum theories agree with all the points of particle search?
- Does the mechanism of transformations of one form into the other exist, and in what forms is it possible?

If we are to assign photon and other entities of energy into the row or the system of particles as it exists now, it should be obvious that we can treat the fields (e.g., Electromagnetic field) as physical bodies with all related consequences. Whatever, when we have described particle(s), it follows the simple common-sense pattern: a particle is described as a fragment (<u>incomplete or isolated piece</u>) or minute quantity of a fragment, relatively small of the smallest possible discrete portion of matter.

In this case, we definitely should look for a multitude of particles of such physical body, and the smashing and splashing do make sense.

Fields can not be treated as a body but rather in the ways they are treated in electrodynamics or a <u>complete and discrete portion of matter</u>.

All that we know about the E-M field, including the interaction <u>between E-M field(s) and particle(s),</u> is well researched and described in quantum mechanics, classic, and to some degree, quantum electrodynamics.

There was no information of any kind about **splitting the E-M field** into fragmented or minute quantities of a matter. The E-M field can be described by its entities of different E-M characteristics, such as polarization, amplitudes, frequencies, etc. A photon, as a basic quantum of E-M energy, can evolve from most particles or their interactions. On the other hand, a photon cannot be described as a particle of a split E-M field.

Hence, all that work that was done in search of idealistic "goddamn particles" is just a waste of resources and time. A recent discovery of "Higgs particle" proved only one thing - when more energy is pumped into the system, the more energy, in all possible forms, will be siphoned out.

That much information was squeezed out of the costly Supercollider experiments.

Quantizing in Electrodynamics

Successes in quantizing the behavior of particles, using the mathematical tools of Quantum Mechanics, has encountered serious problems of inconsistency, when it was applied to the field of Electrodynamics. Classical waves of Electrodynamics stubbornly refuse to be squeezed into the rigidity of mechanics. The E-Field theory can, to some extent, bridge the existing gap of these disagreements.

Based on the E-Field theory, prime energy is a homogeneous entity of continuous function. The process of absorption of this energy by a particle imposes limitations of space boundary on the propagation of the field of Primary Energy. But the following caveat should be considered in the description of the process.

Contrary to the process described by Quantum Mechanics, the process of absorption of the Primary Energy is always (anyplace/anytime) the continuously running process. Hence, tools of the theory of probability, with the well-known Schrödinger's uncertainty principle, are not applicable in this case.

The Primary Energy is absorbed in quantized amounts, but it is done in time and space on the continuous basis. The tools of Quantum Electrodynamics are, more likely, applicable starting from the process of the transformed energy containment within and its emission from a particle.

PART III

NEW MODEL AT WORK

And There Was Light

The Gravitational Model at Work

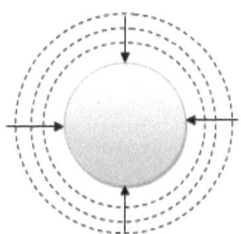

Based on the equations 10.12 and 10.13, a stream of E_{0-} field creates direct flow toward the center of the body. The force of this flow is directly proportional to the mass of the body, its complexity and inversely proportional to the surface of it.

The following derives from our observations of Nature:

- ➢ Small, heavy, and more complex bodies have more gravitational force.
- ➢ Within the same body, gravitational force changes relative to its spherical value of the penetrable surface.
- ➢ Objects with "0" mass have "0" complexity and no gravitational pull.
- ➢ Objects with "0" complexity have "1" for the mass capacity, which may provide for some basic gravitational interaction.

The theory of exchange dictates that transformed energy will be expelled outward from the body and, in doing so, will create a force barrier that prevents possible collision with or absorption of other objects. Only objects with sufficient enough speed can break through this natural barrier.

As a result of the above, we can conclude the following:

1. Based on the E-theory, our planet is surrounded by the incoming and outgoing E-fields and because of that, "some kind of ether wind" cannot exist on its surface. Moreover, since every single body in the Universe has its own shields of E-fields, the process of the wind existence and its detection in any space is out of the question.
2. Based on the Energy Balance and E-field theories, we can say that the amount of incoming or Prime Energy is equal to the sum of the amounts of outgoing or Transformed and Retained Energies:

$$E_0 = E_x + E_r \qquad (20.1).$$

At the same time, the process of E-transformation is not a simultaneous process. It takes a certain amount of time for a substance to absorb, transform, and expel energy. The entire process is a function of time, mass, level of complexity, and speed of an object in relation to the E-field, as well as a density of the E-field in the object's vicinity.

Because the E-transformation process is the function of time and E-field density in any given space, the matter, in any given time, can retain more or less energy than it needed.

Characterizing Space and Time

It is very hard to determine when interpretation of space in time as real objects has begun, but it can be stated with great probability that Newton brought them up to the level of absolute standards. Kip S. Thorne, in his book *Black Holes and Time Warps*, addressed it as follows:

> *Newton's Absolute Space was the space of everyday experience, with its three dimensions: east-west, north-south, up-down. It was obvious from everyday experience that there is only one such space. Newton's Absolute Time was the time of everyday experience, the time that flows inexorably forward as we age, the time measured by high-quality clocks and by the rotation of the Earth and motion of the planets.*

Based on our discovery, we can review an existing description of Space and Time and establish a description of these widely used depictions.

Space

Webster's Dictionary defines space as:

- "That which contains and surrounds all material bodies"
- "Is the idea which each person forms and develops"

From these descriptions, it is obvious that space is not a substance, but an abstract idea devised by humans for only one reason - to make it possible to describe the placement and movement of physical bodies in it. Space has the same meaning as the idea of "sky."

Because it is an abstract idea, it cannot move, warp, twist, be existent, or be measured. Any matter can move in space, be in space, and/or be measured in space. There is no absolute or relative space, just as there cannot be an absolute or relative abstract idea.

Time

Webster's Dictionary gives the following description:

- "The general idea of successive existence, the measure of duration"

If the first part of the definition describes the philosophical point of view, the last part describes "Time" from the technical point of view.

Based on the above, we can conclude that in practical and philosophical terms, time is a measure. It is a subjective, dependent, artificial, and conventional tool that was created by humans to conveniently describe processes that are taking place during observations. Because of the significant body of work done on projecting time as a physical dimension, I would like to take a more in-depth look into this phenomenon.

Historically, the measure of time began with the observation of celestial movements. The most obvious naturally recurrent phenomenon is the alteration of light and dark periods of a "calendar day." The other most important, less frequently recurring phenomenon is the change of seasons.

The basic part of our time-measuring process is in artificial division of the processes of recurrences into usable fractions. The time-measuring tool known as a clock has changed dramatically from its first appearance. The earliest clock was a sundial, which measured the position of the Sun against the Earth. The introduction of a mechanical clock began with the water clock and led to the division of the day into twenty-four hours. The pendulum clock, invented by Huygens in 1656, was a great step that helped to divide hours into smaller units (minutes, seconds, and fractions of a second). A Cesium Fountain Atomic clock is one of the most accurate clocks available today.

In practical terms, when we are observing or recording time, we are observing the measurement of the frequency of two known appearances. The time is determined as a derivative from the frequency of the chosen event:

$$T = \frac{1}{f} \qquad (20.2),$$

Where: T—time,
f —frequency.

At the same time, we are using frequencies to describe cycles.

Webster's Dictionary's definition of *cycle* is "an ordered series of phenomena in which some process is completed."

Replacing "appearance" in our description of time with "cycle" and taking into consideration that only the objective result of any observation is a cycle of any event, we can finally describe time as follows:

> **TIME is a measuring tool that is derived from the measurement of frequencies of chosen Cycles.**

Every act or phenomenon of nature can be described as some sort of Process, which we note as "p." Continuity of a process can be split into individual or discrete parts, which we will call a cycle and note as "i_i" and "t_k." The cycles can be grouped into Σ_i, and after, the groups can be additionally grouped into Σ_k.

Therefore, we can write:

$$p = \sum\sum t_k t_1 \qquad (20.3).$$

Consequently, time is an abstract, an artificially created tool to specifically describe the continuum of a period, cycle, and process. We cannot bend, stretch, or alter time because it is just an abstract measuring tool.

The Big Bang Theory

Kaaa . . . Boooom

I view the Big Bang theory as an attempt of the scientific community to return back to Newtonian absolute time, regardless of obvious discrepancies of this theory with the reality. It is noticeable that the progression of the Big Bang theory is leading us toward another "doomsday" scenario and back to the "Dark Ages" of science.

Contrary to the Big Bang, our model represents the image of our universe that travels within the field of prime energy (i.e., E_0-field). In the following chapters, I am providing alternative points of view that are more in tune with Nature. These chapters demonstrate that there is no *Bang* at the beginning, and certainly, there will be no *Bang* at the end.

Red Shift and other discoveries in Astrophysics resulted in the theory that all matters in the universe were created in the act of violent explosion called the Big Bang. A presently prevailing point of view is that all matters are moving away from one another (see Red Shift Theory). Therefore, by applying the reverse compiling, it would be possible to determine the very time and place of the "Great Beginning." In other words, we would be able to find absolute "0" time and "0" place (welcome back to the "Dark Ages").

Moreover, it was concluded that "0" time began some 5×10^9 to 6×10^9 years ago and since quasars are the farthest objects in space from the Earth, they are the first product of the Big Bang.

Modern science determined that light from the closest quasars travels on average 1.5×10^9 to 1×10^{10} years before reaching the Earth.

The following are simple observations of the above numbers that reveal number of possibilities:

1. Light from quasars should have been emitted before the *Bang*.
2. Having presice current positon of quasers and presently observing light emitted from quasars, due to the constancy of the speed of light in free space, we would be able to determine where quasars were when light was emitted. Unfortunately, we have no idea where they (quasar) are now. On the other hand, applying the Big Bang theory, the quasars at the "Bang time" would have been located exactly where they are right now. It is a bit complicated, but it has to be understood that since light has the finite speed, it took an average of about 10^9 years for the light from quasars to reach us at Earth. By that time, since quasars travel with significant speed and are independent from emitted light and away from us (see Red Shift theory), they will have to move even much farther from us than they are now. Hence, it has to be $10^9 \times 10^n$ years since the Big Bang's *bang*.
3. If the last one is true, we should have observed some results of the reverse process and have to be ready for the collapse of the universe.

Special Relativity vs the Big Bang

Einstein, in his Special Relativity, clearly demonstrated, through the significance of the universal constancy of the speed of light in vacuum, that the principles of absolute time and space (in his words) is just "an abstraction, untenable and artificial."

Big Bang, on the other hand, was conceived as the idea of "Absolute zero time" and "Defined point" in space, which contradicts the principles of relativity. So which theory is real, and which is a fantasy?

Obviously, such unsubstantiated theory was built based on something more fundamental, something that would provide definite and unquestionable support. This something was the Red Shift theory, which we will address next.

Explaining the Red Shift

Even though processes identified as a "Red shift" may be found in micro- and macrophysics, most often the "Red Shift" is attributed to the astronomical phenomena.

The Red Shift in astronomy describes the discovery made by Edwin Powell Hubble, who discovered the frequency shift in the spectrum of the known chemical elements, which arrived as a part of light from an outer Solar System.

At present time, this effect is solely attributed to the Doppler Shift (i.e., changing of the frequencies caused by velocity of recession). Based on this theory, most astronomers regard the Red Shift as the evidence of the process of expansion of our universe.

There are still some interesting questions and thoughts, such as the following:

- ➢ Why is the Red Shift so evident in all measurements and from all directions?
- ➢ If the Red Frequency Shift is the result of light emitted from receding or moving-away bodies in the dynamic system of the universe, why have we not witnessed celestial bodies that are moving toward the Earth?

> Having myriad stars that are rotating inside of myriad of moving galaxies, would it be possible to detect some stars that are moving toward the Earth?

Even Hawkins's model of the balloon-like universe cannot explain that. The assumption made by Hawkins can be applied to the spherically expanding universe of astronomical bodies that are moving at unidirectional trajectories. If we take into account rotational motion of solar systems, galaxies, etc. . . . somewhere, we have to have motion of some planetary bodies toward us that have to produce a blue shift. Since we have not witnessed this phenomenon, we have to look for an alternative model that describes the existence of the Red Shift.

More likely, in the Red Shift phenomenon, we have encountered the solid case of Lorentzian transformations. Since our own spectral measurements were done inside of our frame of references (f_0), the information received from the outer frames (f_1) must be transformed to the measurements of our own frame using

$$f_1 = f_0 \sqrt{1 - \frac{V^2}{c^2}} \qquad (20.4),$$

$$f_1 < f_0 \qquad (20.5).$$

The results of this exercise revealed that the frequencies of the light received from any outer frame moving in any direction and at any speed, "V," should have the **Red Frequency Shift**.

By solving the Red Shift puzzle, we have dismantled the very base upon which the **Big Bang** theory was constructed. In the following chapter, we will discuss how the removal of the Big Bang opens doors for understanding our terrestrial evolutions.

The Process of Sun Burning

In the Subchapter "**Energy Exchange**" of this book, we discussed how interaction between energy and matter creates motion of a matter, which can be described as follows:

> *Having knowledge of a substance behavior inside of an electromagnetic field, we can assume that the process of interaction between the E_0-field and the inner matter energies will create a force that will provide movement to a substance. It will work similar to an eddy current created inside of a piece of metal that is placed into a magnetic field. At the same time, the process of the energy exchange will work as a brake force that prevents the body from constant acceleration. The summary of these two opposite directed forces will establish the speed of motion of matter in the E_0-field. It means that as soon as a body has formed, motion becomes the necessary part of its existence.*

Based on this observation, we can basically explain the existence of hot objects (we call Suns) in the center of any solar system.

The Sun burning is due to the fact that it (a planet like body) is located at the Center of a system and as such, it has restricted spatial motion. The imbalance between the amount of energy, retained by such body and the motion that is required by the energy of the exchange process, creates an excess of energy, which is the base of "Burning."

Since the process is the function of E_0-field density, we have different amounts of "burning" of the same matter, which we call Sun in any given time.

That difference resulted in the physical processes that we are observing on the surface of the Sun and the amount of energy it emits toward surrounding planets, producing thermal imbalances in the solar system.

Description of Drastic Changes of the Earth's Climate

The traditional Big Bang model describes the progression of the Earth's development in the *"From burning hot to cooling down"* sequence. This model describes a post Big Bang object of molten lava that slowly cools down to the solid-crust state. Water and primitive life were apparently brought to the Earth by meteorites and other spatial objects that fell on our planet during the initial phases of its development.

Here are some inconstancies. If everything has developed at Big Bang and everything in the universe was hotly hot,

- ➢ why, by all admission of science, are all solid objects in the universe aging differently?
- ➢ where, when, and how was ice formed on the stellar bodies, including meteorites?
- ➢ why has ice been delivered by meteorites?
- ➢ where, when, and how was life formed in the extreme heat of the universe?
- ➢ where, how, and from what was water formed?

And finally, the most significant question is, how can it explain the drastic changes between the warm Jurassic period, Ice Age, and rewarming of our planet?

Here is how our new Energy Exchange Model can explain most or some of that.

The center of our solar system - the Sun (as many other Suns in the Universe) - burns some extra energy. In the subchapter "The Process of Sun Burning," I have described the fundamentals behind the "Solar" process, As I have pointed out, the amount of energy of any system, including solar systems, is a time-related function of the density of E_0-field energy in a given space and time. Hence, the Sun burning that produces released energy is proportional to the density of the E_0-field in given space and time. More E_0-field density corresponds to a more active Sun and hotter Earth; less density corresponds to a less active Sun and cooler Earth. The output of the Sun's energy provides for changes in the Earth's climate, making it tropically hot or polar ice-cold.

The life cycle in the Earth is changing in accordance with climate conditions and the influence of the density of the E_0-field. That is the only phenomena that can be credited for such drastic changes in the Earth's geography and biohistory. Paleontology identified significant changes in the life of past geological periods. There is number of hypotheses regarding the reasons for such changes. The prevailing theory is built around the idea of meteorite collisions with the Earth, which created a dust cloud and extended winter conditions that instantaneously, in a flash, destroyed tropic-like life and killed dinosaur-size animals.

Our theory provides the most plausible explanation of prolonged periods of low E_0-field energy that correlates with the Ice age and the high E_0-field energy of tropic-like conditions.

That means that the Earth may get into prolonged ice or tropic-like conditions in the future.

Black Hole Theory in the View of Our Exchange Energy Model

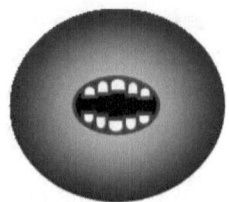

The most modern Black Hole Theory describes the phenomenon of nature as a matter-eating monster, which uncontrollably absorbs stars, solar systems, and all. The center of the Black Hole is considered to be a superdense matter that as the theory describes, is nothing more than the same matter but which collapses on itself. If this is correct, the modern monster called "Black Hole" will have no natural halts, which will preclude it from swallowing everything in its neighborhood and beyond (in electronics, it is called "total positive feedback of a runaway" system). Having such a monster at the center of our galaxy means that sooner or later, it will absorb everything, including our solar system. Such an ugly prospect will give us even less time to exist than the idea of the burned-down Sun.

From the point of the E-field theory, the Black Hole can be described as a place without an E_0-field. The lack of an E_0-field will create the phenomenon that is described as the Black Hole, except for a superdense center. The absence of an E_0-field can be produced by the following conditions:

> ➤ Absolute vacuum is the natural effect. It is right to assume that the E_0-field is not everywhere in the universe. More likely, it has its normal boundaries.

> The E_0-field deficit can be produced by number of other phenomena.

One of these phenomena can be described as space constraints. The process of the Energy Exchange requires that the certain amount of E_0-field will be available for the process of transformation, which is constantly needed for support of matters. Under such conditions, the center of a galaxy can be so crowded that the amount of space needed to provide the required concentration of the E_0-field will not be sufficient to support such an amount.

In that case, the available E_0-field will be distorted or drawn toward the matter, and therefore, the very center of the system (galaxy) will appear to be partially or totally empty of E_0-field.

The level of emptiness will not be constant, but most likely, because E_0-field is not consistent, it will be a function of the density of E_0-field in that particular area. The change in transparency or "blackness" of a "Hole" may be observed with time.

The "Parker Effect"

The "Parker Effect" and Contemporary Relativity

The National Aerospace & Electronic Conference NAECON was the place where the Parker Effect and related work on "Velometer" and the theory of Non-Inertial Navigaiton were reveiled to the public in 1997.

The "Parker Effect" takes its roots in Maxwell's Electrodynamics and utilizes the same theoretical and mathematical apparatus that were used by Einstein in his Special Relativity and by Lorentz in the Transformation.

At the beginning of the twentieth Century, Lorentz provided the tool by which measurements made in different inertial frames can be compared with the speed of light and with one another.

As we have mentioned before, this tool is $(1 - u^2/c^2)^{1/2}$—which we called Lorentzian correlator.

However, whereas Lorentz focused on determining how Electrodynamics influences Galileo's transformations, in the case of the "Parker Effect," we concentrate on investigating the correlation between inertial and noninertial media. The results we have acquired from this work have laid the foundation for Contemporary Relativity. Based on the "Parker Effect," Contemporary Relativity rejects the notion that all constant velocity motions are indiscernible from one another or from relatively stationary states. It establishes the ways to detect constant velocity motion from within of an object in motion and provides ways to define the characteristics of this motion. In doing so, the "Parker Effect" simplifies and streamlines Relativity, unifying it with the laws of Electrodynamics.

We will begin our analysis with Einstein's two postulates and Maxwell's equation for uniform plane electromagnetic wave propagation in free space.

We use Maxwell's equation for uniform plane waves in free space, where permeability of free space $\mu_0 = 1$, permittivity of free space $\varepsilon_0 = 1$, and attenuation of free space $\sigma = 0$, since:

$$B = \mu_0 H \text{ in free space} \quad (20.6),$$

$$\frac{E}{B} = \frac{1}{\sqrt{\varepsilon_0 \mu_0}} = c \quad (20.7), \text{ or}$$

$$E = Bc \quad (20.8).$$

The E and H vectors in free space are in phase because the characteristic impedance of free space is real. Since the two vectors are orthogonal and in phase, the E × H pointing vector is a straight line pointing in the direction of the wave propagation.

From Einstein's First Postulate, we can conclude that the light inside any inertial system will obey the laws of electrodynamics.

From Einstein's Second Postulate, we can conclude that a pulse of light will continue its propagation with a speed that is independent of motion of its source.

From Maxwell's equations for uniform plane waves, we can conclude that a pulse of light emitted into free space will hold its direction of travel independent of the motion of its source.

Therefore, summarizing the above, we can state the following:

> **The Postulates of Electromagnetic Wave Propagation in Free Space**
>
> 1. From the moment when a beam of light is emitted, the motion of an inertial system from which it was emitted (its source) becomes external to the propagation of the light.
> 2. Light does not inherit any of the inertial characteristics of the motion of its source.
> 3. The direction that the light will travel in free space is only the function of the aim of its emitter, not a direction of the emitter's travel.

Based on the postulates of Electromagnetic Wave Propagation in Free Space, the "Parker Effect" states the following:

> **The Parker Effect**
>
> A pulse of light due to its noninertial nature will hold its constant trajectory and speed in free space inside of an object (inertial frame) in motion while each member of the object, including the emitter/receiver system, will continue its movement. This result is valid for any type of velocity and motions in space.
>
> The Effect is that we can derive the data on a vector of relative velocity, position, and orientation of an object (inertial frame) in 3-D space by comparing concurrently present and past data on the two correlated systems (inertial and noninertial) at any given time.

Displayed below is Fig. 20.1. It demonstrates our new method of detecting the velocity of an object's motion from within and from the motion itself.

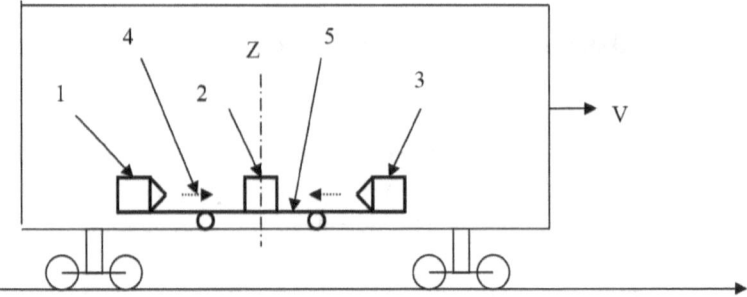

Fig. 20.1

The platform 5, which is comprised of the light emitters 1 and 3 and is equally spaced along the X-axis from the interferometer 2, can freely rotate around the Z-axis. At certain intervals of time, pulses of light 4 are simultaneously emitted from the light emitters 1 and 3 toward the interferometer 2. The output from the interferometer, when the platform is in the position as shown, can be compared with the results obtained as the platform rotates around the Z-axis. It is obvious that the difference between the initial and subsequent readings will be greatest when the platform will be oriented along the Y-axis or perpendicular to the initial setting. The information from the comparison of the two readings will establish relative velocity of the boxcar in the X-Y plane.

The physical model of this method can be represented by the 3-D light box shown in Fig. 20.3.

It seems that Einstein himself established our method when he devised the following Thought Experiment to illustrate the simultaneity and relativity of time (Fig. 20.2), where he demonstrated that inside a moving boxcar, the light from the front of the boxcar will reach the observer faster than the light from the rear.

The observer inside the boxcar will therefore conclude that event B' occurred before event A'.

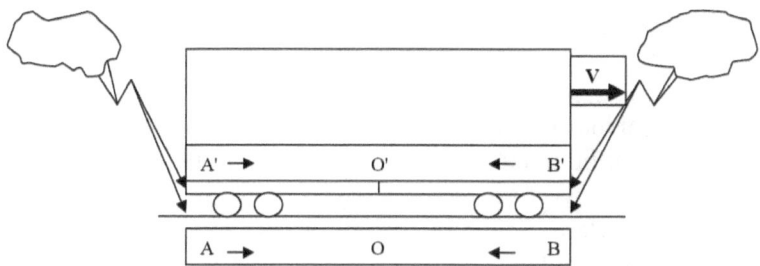

Fig. 20.2

In this experiment, Einstein uses his postulates to demonstrate that the light emitted inside an object in motion is unaffected by the boxcar's motion. It is obvious that the asynchronous arrival of equally spaced and synchronously generated light pulses can only be attributed to the influence of the velocity of the boxcar, which is correlative with the trajectory of the light.

Theoretical Light Box

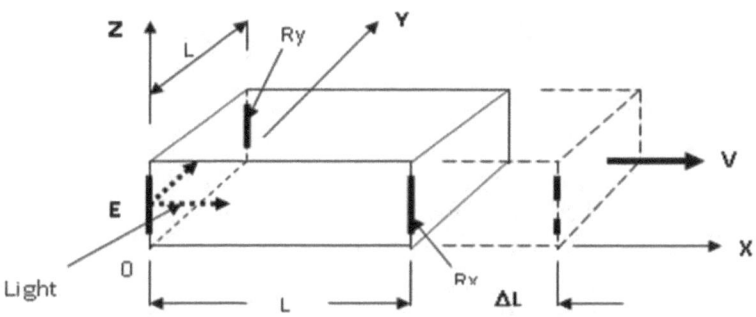

Fig. 20.3

The FIG. 20.3 represents a 3-D square light box, which consists of light emitter, "E," and two receivers, "Rx" and "Ry," placed orthogonally in the opposite corners and at an equal distance, "L," from the emitter. The light box travels along the "X" axis with a constant velocity, "V," and with respect to the inertial frame of references.

At particular intervals of time, the emitter, "E," emits simultaneously two pulses of light: one pulse is emitted toward the receiver "Rx" and the other toward the receiver "Ry." In this experiment, we will measure the elapsed time of arrival for both pulses and compare them with each other. The first pulse of light is moving between the emitter, "E," and receiver "Rx" along the "X" axis in the direction of the light box's travel. The other is moving between the emitter, "E," and receiver "Ry" along the "Y" axis or orthogonal to the direction of the light box's travel.

At time, "t_o," the two orthogonal pulses of light were simultaneously emitted. Since our light box is moving at a constant velocity along the "X" axis, by the time the pulse of light, which is traveling along the "X" axis, will reach receiver "Rx," the light box will move a distance, "DL." The elapsed time in this case can be determined as follows:

$$t_x = \frac{L + \Delta L}{c} = \frac{\Delta L}{V} \qquad (20.9),$$

Where

t_x—elapsed time for the pulse in "X" direction,
L—distance between emitter and receivers,
c—speed of light,
V—velocity of the light box.

On the other hand, the pulse of light, which moves parallel to the "Y" axis and transversely to the vector, "V," will reach the receiver Ry at time:

$$t_y = \frac{L}{c} \qquad (20.10).$$

Obviously, we can state that in our case,

$$t_x > t_y \qquad (20.11).$$

In this case, when the light pulse will travel in the opposite direction to the direction of travel of the light box,

$$t_x < t_y \qquad (20.12).$$

In general terms, for any moving objects, we can conclude that "t_x" can be larger or smaller but never equal to "t_y."

By using our method, we can solve the three-hundred-year-old problem that was first introduced by Galileo and later modified by Einstein. The problem was described as follows:

- If a passenger boards a train (or a ship in Galileo's portrayal), can he/she tell a difference between a stationary (relatively to the ground) position of the train and a constant velocity motion?
- Is it the train that moves away from the station or the station moves away from the train?

Based on the results produced by our method, the passenger from within the train can establish the train motion relative to the ground and to the station.

From a Light Box to the Velometer

Uniqueness in the discovery of a light box can be easily applied to the development of the devices that are measuring the velocity (speed and direction of motion) of any object from within and from the motion itself. I named such type of device "Velometer" to distinguish it from a Velocimeter, which measures velocities of outside motions.

Since light has no preferences in its direction of motion, Velometers can utilize measurements of light that propagates collinearly with the motion of the object or transversely to it.

The collinear type of the Velometer can be described by the following set of equations using equations 20.9 and 20.10:

And since $\quad\quad\quad \Delta t = t_x - t_y \quad\quad\quad (20.13),$

From the equation 20.14

$$t_x = \frac{L}{c} + \frac{\Delta L}{c} \quad\quad (20.14),$$

Substituting

$$t_y = \frac{L}{c}; \Delta L = t_x V$$

$$V = \frac{c\Delta t}{t_x} \qquad (20.15).$$

Equation 20.15 demonstrates that the measurement of the object's velocity along the X-axis can be found from the measurement of the times of arrival of the beams of light at the Rx and Ry receivers.

Having the light receivers placed along all three axes will provide relative measurements of the object's velocity in three-dimensional space.

The Collinear type of the Velometer can find its applications in a variety of cases, especially where the speed of an apparatus is significantly less than the speed of light. In this case, Fitzgerald's space contraction that acts in the direction of travel will not measurably influence our measurements.

At speeds where Fitzgerald's space contraction can influence measurements, the transverse model will yield more reliable results. Referring now to Fig. 20.4 and Fig. 20.5, the horizontal axes in the views represent the distance. The Postulates of Electromagnetic wave propagation state that light does not inherit the inertia of its source; hence, the light pulse will travel in a straight line. At time, "t_o" (Fig. 20.4) a pulse of light (4) is emitted from the emitter (2). At the time, "tn" (Fig. 20.5), the pulse will reach the array of receivers (3), but the device (1), at that time, have moved in a distance, "L," and will be located at the new position, hense, light will arrive to the different element of the recever.

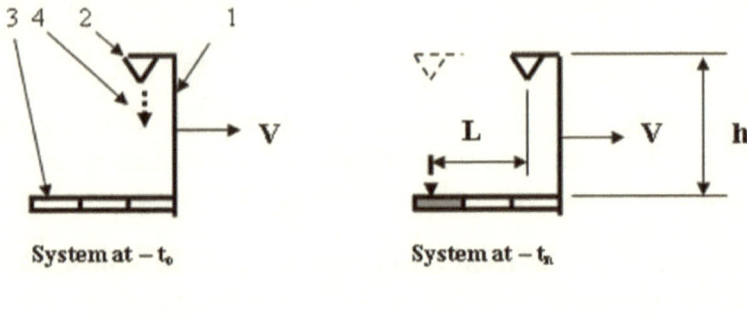

Fig. 20.4 Fig. 20.5

The movement of the device (1) and the pulse of light (4) occurred during the same period of time, "t_i," which is equal to

$$t_i = t_n - t_0 \qquad (20.16).$$

We can write the following equation as

$$t_i = \frac{h}{c} = \frac{L}{V} \qquad (20.17),$$

Where

    c—speed of light,
    h—distance between emitter and receiver,
    V—speed of device,
    L—distance the device has traveled at the time, "t_i."

Since "c" is the universal constant and "h" is the built-in constant for a particular device, we can write the following:

$$\frac{c}{h} = k \text{ (constant)} \qquad (20.18),$$

$$V = kL_1 \qquad (20.19).$$

The last equation (20.19) describes that the displacement of the device (1), in relation to the trajectory of the pulse of light (4), is proportional to the speed, "V," with which the device is moving, and the direction of this displacement is the direction of travel of the device in space.

Once again, I would like to point out that our method will provide only relative measurements and measuring devices that are built on its principles that can be calibrated in relative terms only.

The Velometer—the Device of Noninertial Navigation

Among many discoveries, there are some that live through the centuries. And among those, there are only a few that share their destiny with the destiny of human race itself. In my opinion, Velometer is such a discovery. Since Velometers provide navigational information by utilizing the uniqueness of the relation between the inertial bodies and the propagation of noninertial media (light) within, they are the only devices that will guaranty reliable and independent navigation even in the deep-space applications.

It is my gift to humanity that will serve as the device of Noninertial Navigation, guiding humans in their eternal quest for space exploration and ultimately in the strive for survival.

> **Motion in space, regardless of its nature, is detectable and measurable from within the system and from the motion itself.**

Space Telescopes or "What We Have Not Learned from Troubles with the Hubble"

The most immediate practical application of the "Parker Effect" can be found in the design of space-based platforms. On the other hand, neglecting this effect in the development of space-based equipment as we have discovered costs a significant amount of money.

Back in December of 1997, the *Photonics Spectra* magazine published the article "Point the Way to a Stellar Alignment," which described alignment procedures done by Eastman Kodak and TRW Inc. on a space-based x-ray imaging observatory. In this article, Charlie Atkinson described that the optical system was aligned at the Eastman Kodak facility at a distance of 723.9 mm above the optical bench and needed 0.2 mm further realignment when it was placed in TRW's final assembly tower at a distance of 8.8 m. I have contacted NASA, Eastman Kodak, and TRW staff to warn them about the possibilities of troubles that the space-based telescopes may incur after deployment. After my warning was neglected, I wrote the article that was published in the July 1998 issue of the *IEEE AES Systems Magazine* as cited below:

"What We Have Not Learned from The Troubles with The Hubble"

The Hubble space-based telescope is a great tribute to our progress in space. The abilities to place an optical telescope at a significant distance from the Earth's surface, away from the interference of the planet's unsteady atmosphere, have already paid off by producing magnificent records of the astronomical activities in depths of outer space.

Nonetheless, for all glorious information the Hubble telescope has provided, we cannot forget its rough and troublesome beginning. To these days, the problems with the alignment of the Hubble's optics were blamed on the manufacturers of its optical components. The hastily set investigation concluded that the problem is a spherical aberration of the primary mirror (the primary mirror is said to be 2 microns too flat at the edges).

Since the time of Galileo Galilee, all telescopes were built, aligned, and used on the Earth's surface. The Hubble is the first telescope to be built and aligned on the Earth but used in space. Because of this, we have to consider the fundamental difference between the alignment of surface-based and space-based telescopes.

For those who missed our article "The 'Parker Effect' and Navigation in Space," The "Parker Effect" describes the result of the interaction between inertial bodies (anything that has mass) and non-inertial media (light or other E/M fields).

The "Parker Effect" and the Optical Alignment of the Space-Based Telescopes

It is a well-recorded scientific fact that a beam of light (Noninertial Media, NIM for short) will remain on its trajectory (the direction it was emitted to) regardless of the movement of any inertial system (physical body) that the NIM is traveling through, within, or in the vicinity of.

We will describe the "Parker Effect" and its effect on the space-based *x-ray* observatory with the help of Fig. 20.6, Fig. 20.7, and Fig. 20.8.

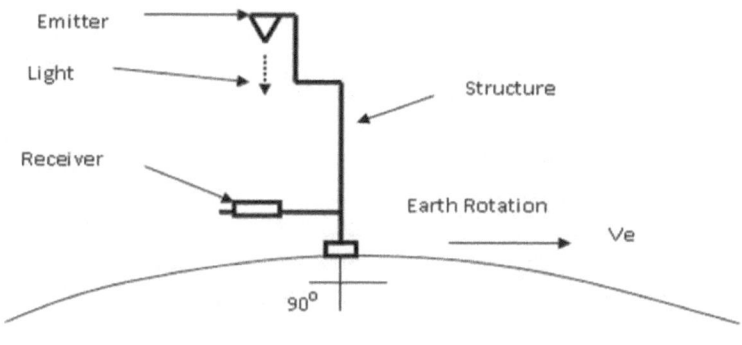

Fig. 20.6

Fig. 20.6 shows a simplified view of the space-based telescope's optics alignment structure and its relation to the orbital rotation of the Earth.

As it is reflected in Fig. 20.6, the optics alignment structure is anchored to the ground and leveled in its vertical position. Since the Earth rotates with a linear velocity, Ve, around its vertical axis, with it are the alignment structure, the Hubble telescope, and all the related equipment.

To simplify our demonstration, we replace the complex optical path of the real fixture with a straight-line path demonstration (such a simplification will not influence the results of our findings).

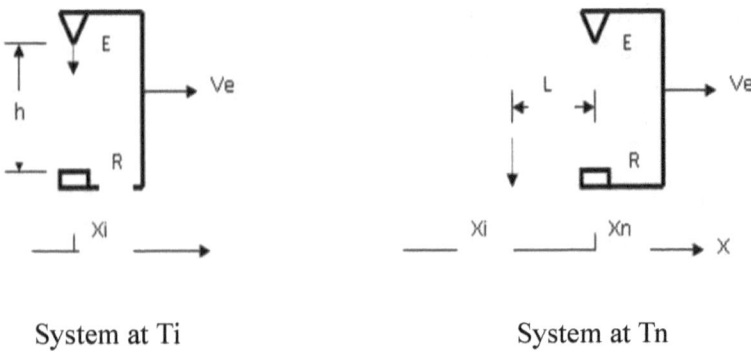

System at Ti   System at Tn

Fig. 20.7 Fig. 20.8

Fig. 20.7 shows the ray path in the alignment structure, which is taken at Ti -the time when a pulse of light was emitted. Fig. 20.8 shows the same ray path at the time, Tn, when the pulse of light reaches the receiver's level. In using Fig. 20.7 and Fig. 20.8, we observe the travel of a pulse of light inside the inertial system at different frames of time.

The alignment structure consists of the emitter (E) spaced vertically at a distance (h) from the receiver (R). Since all parts of the structure are physically connected to one another and altogether to the ground, the structure is moving with the earth's velocity (Ve) in space.

At time Ti, a pulse of light (s) is emitted from the emitter (E) toward the receiver (R). Because light is a noninertial media, it will keep its trajectory regardless of the movement of the inertial system. And by the time Tn, when the pulse of light will arrive at the receiver's level, the whole system will have moved from position Xi to position Xn or a total distance (L).

Using The "Parker Effect," we can establish the distance of the displacement (L) of the inertial system in relation to trajectory of the pulse of light:

$$\frac{c}{h} = \frac{V_e}{L} \quad \text{or} \quad L = \frac{V_e \times h}{c} \qquad (20.20),$$

Where

"c"—speed of light,
"h"—distance between emitter (E) and receiver (R),
"Ve"—linear velocity of the Earth's orbital rotation,
"L"—pulse displacement.

Let's consider the specifics of the Hubble telescope, where

- "h" is the distance between the primary and secondary mirrors equals to 4.88 m (app.),
- "Ve" is the earth's known velocity at the level of New York equals to 298 m/sec (app.),
- "Vs" is the velocity of the telescope in orbit 600 km above the Earth's surface equals to 461 m/sec (app.), and
- "c" is the speed of light equals to 300,000,000 m/sec.

Applying the listed above to the specificity of the Hubble telescope data, we can compute the simplified optical system misalignment as follows:

- On the Earth's surface

$$L = \frac{298 \times 4.88}{300{,}000{,}000} = .00000485$$ m or for the Single Path System (as is shown in the Fig. 20.8 L = 4.85 mkm).

- In the orbit

$$L = \frac{461 \times 4.88}{300{,}000{,}000} = .000007498$$ m or for the Single Path System (as is shown in the Fig. 20.5) L = 7.5 mkm).

The comparative measurable difference on misalignment due to the "Parker Effect" for the Hubble telescope is **7.5 mkm − 4.85 mkm = 2.7 mkm.**

It is obvious, since the real optical path in the observatory is more complex than what was shown in Fig. 20.7 and Fig. 20.8, that the calculated misalignment of 2.7 mkm is well within the range of 2 mkm. To these days, the problems with the alignment of the Hubble's optics were pinned on the manufacturers as a result of the spherical aberrations.

The most significant information that is pointing in favor of the "Parker Effect" can be found in the Government report on the analyses of the troubles with the Hubble telescope, the picture that is displayed below (Fig. 20.9). This set of four pictures from the report provides the critical link between the optical troubles with the Hubble Telescope and the "Parker Effect."

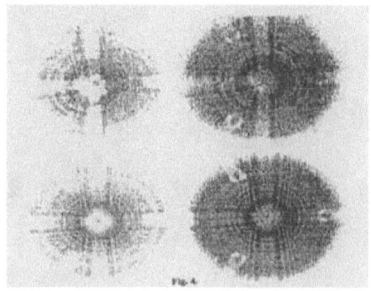

Fig. 20.9

The picture is a Courtesy of Perkin-Elmer Corporation.

As it can be clearly seen from the computer simulations, a spherical aberration (Fig. 20.9, bottom two images) would produce a clear concentric pattern on the images. On the other hand, the two images provided from the Hubble telescope (Fig. 20.9, top two images) have a clearly visible nonconcentric tilt and even an off-center overlap. The result of such patterns can be produced by the angular misalignment between the two mirrors. The angular misalignment, however, will also produce stretched or oval-shaped images. Since all the images are circular, the axial shift is only possible if one image travels parallel to the other in space, and that is possible only under the "Parker Effect."

> An optical telescope or any other apparatus designed to be used in space-bound conditions, which uses noninertial media (NIM) as part of its function, has to be aligned and calibrated using correlators provided by The Parker Effect.

The Velometer Device as a Planetary Compass

Traditionally, traveling on a planet (e.g., Traveling on the Earth) requires some kind of guidance references. Historically, Earth explorers were guided by the terrestrial observation of surroundings, the observation of a night sky, and the nature of the Earth's magnetic field. These methods require a collection of longtime observations that may not be available on the distant planets.
The modern methods of navigation on the Earth includes ground- and space-based radio frequency referencing that will not be available on the other planets as well.

Using the Velometer device as the ultimate Compass will solve this problem.

Because all planetary bodies are in the semispherical form, each point of the planetary surface arc is moving with different axial speed. Similarly, to the Earth's navigation, travelers can establish longitudinal and latitudinal correlation by measuring the axial speed of the planet.

Since the Velometer device measures velocity of the motion of any object from within and from the motion itself, it would be relatively simple to use the Velometer as a Planetary Compass (Fig. 20.10).

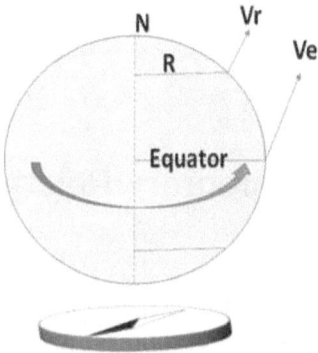

Fig. 20.10
The Universal Planetary Compass though the arc velocities of a Terrestrial semispherical body

Moreover, this method can be used as an ideal tool of navigation on Earth, especially for the development of Autonomous means of transportation.

Traditional Ways of navigating on Earth consists of the referencing of its magnetic field, which has extremely complicated pattern, and/or utilizing outside referencing. Currently, outside referencing includes the system of Ground Control, GPS, and Localized Structures of Cell Towers.

Unfortunately, all these methods, and even the combination of each, have its natural and man-made limitations. Natural limitations include weather conditions and the instability of the signal due to the location (e.g., the areas of tall structures, inside of a tunnel, under bridges, or even in the area that is far removed from accessibility of the signal). Man-made limitations may include the areas of strong Electromagnetic interference. Navigation in such areas is extremely difficult and sometimes even dangerous for a journeyer(s) or manned crafts and practically impossible for unmanned, autonomous crafts.

Since spatial location, position, and orientation are the keys for autonomy, the Velometer-based devices will provide this information reliably and in real time.

Contemporary Relativity

The Present Relativity is a combination of number of separate ideas that, over the years, were interwoven into one joint theory.

The substance of relativity is based on the foundation of the very meaning of the word "*Relative*," which means "something considered in reference to something else."

This term underlines the concept that everything in space is measured in relative terms only, which is contrary to the term "absolute." Since everything in space is in constant motion, there cannot be an absolute space, velocity, time, etc.

Einstein in his "Special Relativity" has clearly demonstrated, through the significance of the universal constancy of the speed of light in a vacuum, that the principles of absolute time and space are just an abstraction, untenable and artificial.

Based on Maxwell's electrodynamics, Lorentz in 1904 extended Galileo's transformations, bringing them in accordance with the constancy of the speed of light and its relation to inertial frames of reference.

During the same period of time in 1904, J. H. Poincaré set forth the following "Principle of Relativity" that became inextricably linked to the discussions on relativity:

> *The laws of physical phenomena must be the same for a fixed observer as for an observer who has a uniform motion of translation relative to him so that we have*

not and cannot possibly have any means of discovering whether we are being carded along in such motion.

It is important to understand that this is an entirely separate concept from the original premise of relativity. Whereas relativity simply states that no measurements are absolute in nature, Poincaré's relativity states that all measurements, taken at rest or in motion, would be indiscernible from one another.

The most common example of Poincaré's idea is a passenger on the train who decided to shut down the blinds. At that, the passenger has no way of knowing if the train is in motion or still at rest without looking outside.

A second part to the above example continues further to claim that if the passenger did look out the window and saw another train, it would be impossible to distinguish which train is in motion. This result, however, has created number of contradictions in modem physics. It would imply that in performing Lorentzian transformation, one can arbitrarily choose which frame (S or S') is to be considered relatively stationary. That assumption will lead us to some kind of mutation of the Twin Paradox.

Using the Velometer device, the passengers can determine their own relative velocity, and therefore, they can also determine the relative velocity of the other train. Likewise, the astronaut would have no question as to his relative velocity. The implications of this discovery go beyond the solution of the Twin Paradox conundrum; most importantly, it will allow astronauts to navigate in Deep Space without external references, which would not otherwise be possible.

The "Parker Effect," therefore, broadens the scope of application of the Lorentz transformations and solves many long-standing contradictions. As a result, the Contemporary Relativity uses the "Parker Effect" to simplify and streamline the conventional relativity by unifying it with the laws of electrodynamics.

> **Since we can determine the motion, including the constant velocity motion of any object from within, the principles of the Contemporary Relativity are based exclusively on the fact that everything in space is in motion. Therefore, it is measured in relative terms only and there are no absolute space, velocity, or time.**

PART IV

ON KINETIC AND POTENTIAL ENERGIES

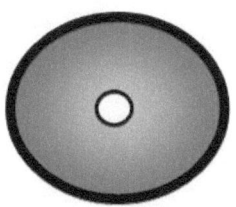

There is Light of Knowledge at the End of a Tunnel

The Contemporary Views on Kinetic and Potential Energies

From the early school days till later in college studies, we took the definition of Kinetic Energy as a postulate of physics. Whether in school, in books, or in scientific literature, the KINETIC ENERGY is described as an object's internal energy possessed by the object by virtue of its motion. Moreover, it is responsible for the relativistic increase of the mass of the body:

$$KE = \frac{1}{2}mv^2 \qquad (30.1).$$

The following is an example of how it is most commonly described in literature, such as *Phenomenal Physics* by Clifford Swartz, The University of New York at Stony Brook:

> *Considering the current importance of energy, it is surprising that the concept was not understood until the middle of the nineteenth century. For all their sophistication, Galileo, Newton, and Franklin did not know that a quantity called energy could be defined in such a way that it is always conserved. Perhaps they didn't discover the idea because the concept is not at all obvious. Energy appears in many different forms. Energy in all these different forms can be defined in such a way that the total energy is conserved as one system transforms into another. However, so long as a system never changes in any way, its energy content is meaningless. It is only during transformations, from one shape to another of from one place to another, that*

the concept of energy becomes useful as a bookkeeping device.

The questions are as follows:

- Where did this concept of energy come from?
- What were used as tools to prove such energy existed?

Three hundred years ago, scientists argued about who was right.

Scientists in Germany championed the preeminence of "mv^2" that was called "Vis viva" or living force. English scientists, on the other hand, were sure that the quality of motion was best described in terms of the momentum, "mv."

As we know today, the German's point of view, unfortunately, has prevailed.

In the subchapter "The Work Necessary to Produce Kinetic Energy" of the same book *Phenomenal Physics*, the following example of the origin of Kinetic Energy was made:

How does an object get to have kinetic energy? The energy can be transferred to the object in a collision, but it can also be produced by shoving the object until it has the required velocity.

Starting from rest, an object will reach the velocity, v, if it is subject of a constant acceleration, a, over a distance, x, satisfying the condition $v2=2ax$. The constant acceleration is produced by a constant force, given by $a = \frac{F}{m}$. If we substitute this formula for acceleration into the formula, for final velocity, we get

$$v^2 = 2\frac{F}{m}x \qquad (30.2)$$

$$\frac{1}{2}mv^2 = Fx \qquad (30.3)$$

If we exert a constant force on an object for a distance, the kinetic energy produced is given by the above formula. The product of force and the distance, through which the force is exerted, is called - work.

Certainly, the maximum acceleration, and hence the maximum velocity, will be produced if the force is applied in the direction of the displacement. In fact, any force, applied perpendicular to the displacement produces no useful effect at all - unless it reduces friction. In the case of the product of force and a lever arm to produce torque, the situation is exactly the opposite. Only the component of the force perpendicular to the lever arm is effective. Apparently, we need another kind of product of two vectors. This one is called the dot product, or scalar product.

$$Work = W = F \times x = [F] \times [x] \times \cos Q \qquad (30.4)$$

The magnitude of the force component in the direction of the displacement is given by [F] cos Q, were Q is the angle between F and x.

The vector product of two vectors yields another vector, perpendicular to the first two. The dot product of two vectors yields a scalar. Work and energy have no direction properties.

According to this definition, no work would be done on a heavy box if you carried it in your arms across the room. The force that you exert to hold it is directed upward, but the displacement is horizontal. Therefore, the angle between force and displacement is 90^0, and the work done on the box is zero. Can this be right?

Even if no work is done on the heavy box as you carry it across the room, surely you do the work. If you had to do that work all day, you would get tired. Your muscles are doing a great deal of work according to the standard definition, just trying to maintain a constant force, although the net displacement is zero.

The unit of kinetic energy is Joule and, since work can produce kinetic energy, we will use the Joule as the unit for work.

Now let's reconsider that ancient question about whether the true quantity of motion is best described by mv or $1/2mv^2$, by moment or by vis viva. If you exert a force, F, on an object for a time interval, dt, then you have exerted an impulse that produce a change in momentum of the object:

$$Fdt = d(mv) \qquad (30.5)$$

If you exert a force, F, on an object, through a distance interval dx, then you have done work on the object. If all the work goes into changing the kinetic energy"

$$Fdx = d\frac{1}{2}mv^2 \qquad (30.6)$$

The question is:
How developed was this measurement?

1. From the definition of *average velocity*,

$$x_t = x_0 - vt \qquad (30.7),$$

which is under the uniform acceleration

$$v_t = v_0 + at \qquad (30.8),$$

and if velocity is a linear function of time,

$$v_{avg} = \frac{1}{2}(vt - v_0) = \frac{1}{2}(v_0 - at + v_0) = v_0 + \frac{at}{2} \qquad (30.9).$$

2. Substituting 30.7 in 30.9,

$$x_t = x_0 + (v_0 + \frac{at}{2})t = x_0 + v_0 t + \frac{1}{2} \qquad (30.10).$$

3. Interpreting 30.8 as follows,

$$t = \frac{v_t - v_0}{a} \qquad (30.11)$$

and substituting it as "t" in 30.10,

$$x_t = x_0 - \frac{1}{2}a(v_t^2 - v_0^2) \qquad (30.12)$$

or using displacement $s = x_t - x_0$

$$2as = v_t^2 - v_0^2 \qquad (30.13).$$

Equation 30.13 shows the relation between the uniform linear acceleration and the average velocity.

4. The following steps were taken to develop a mathematical tool to measure Kinetic or Potential Energies.

The development of the tool starts from Newton's Second Law,

$$F = ma \qquad (30.14).$$

When applied with the following interpretation of speed and acceleration under the application in condition of linear and uniform motion,

$$F = \frac{v_t^2 - v_0^2}{2s} \qquad (30.15) \quad \text{or}$$

$$Fs = \frac{1}{2}v_t^2 - \frac{1}{2}v_0^2 \qquad (30.16).$$

The right side is usually described as the body's change in kinetic energy.

What is describing equation 30.16 in reality?

First, we have to remember that we have constructed this equation from the following components:

- The 1/2 came from the process of averaging.
- v_t^2 or v_o^2 came from application of acceleration over the distance

$$ax = \frac{vx}{t} = \frac{x^2}{t^2} = v^2 \qquad (30.17).$$

In our traditional measuring system, the amount of force delivered to move an object over a distance is called **Work**.

Hence, we may describe the process in the following terms:

The amount of Work required by an Outside Force to move an object at the distance "x" from place "0" to place "t" is equal to the amount of Work to move the object to place "t" minus the amount of Work to move the object to place "0" divided by 2.

There are no accounts of any changes in the object's internal energy that can be directly attributed to the application of the outside force, and therefore, there are no changes in the inside energy.

> Based on the current level of knowledge, we can state the following:
>
> - Matter (inertial physical body) can store or release some amount of energy through changing in or rearranging of its internal conditions, which could eventually lead to rearranging the object's inner structure.
> - There are no other forms or methods, known to science, of storing or releasing energy from any kinds of matter.

The Pendulum Effect was usually cited in the case of demonstration of the energy transformation.

By the traditional theory, once a pendulum is brought to an off-center position, it has accumulated potential energy that, in the time of release, will instantaneously convert itself into kinetic energy. On the other side of the swing, kinetic energy will transfer itself into potential energy and so on. To provide for a balance of energy, as it is required by the law of physics, mathematically potential energy is displayed as a negative number. Therefore, a ball cannot gain energy by swinging back and forth.

Annotation: energy is a **scalar number.**

By utilizing force and momentum, we can eliminate the nonsense of kinetic/potential energies.

Let us use it in conjunction with Newton's third law as follows: a pendulum is in the "0" position, where the force of gravity, which draws the ball toward the Earth, is defying it through the strength of the cable that connects the pendulum to the pivoting point.

By applying our force (in expense of our internal energy), the pendulum will move from the center into a release position. The force of gravity and our force are balanced at the point of release. Once we remove our force, the force of gravity pulls the pendulum toward the "0" position. On its way down, the force of gravity and the cable creates a momentum of motion, which is carried by the ball passing its "0" point and into the other side. At the point where the force of gravity prevailed over the momentum, the reverse swing takes place. There is no energy added to the ball, no energy subtracted, and no energy transformation either.

The same conclusion can be drawn for proving the point against the term of Potential Energy by substituting a linear displacement with height or a vertical displacement.

The very interesting point of observation that is related to our discussion was brought in Newton's *Principia*. It was pertained to the motion of projectiles shot with a certain initial velocity through a resistant medium, like air or water. How far will they move before coming to rest?

While moving through the medium, the projectile must push aside the medium in order to make a tunnel to move forward. Newton showed that the length of the tunnel stands in the same ratio to the length of a projectile as the density of the projectile to the density of the medium:

$$\frac{L}{1} = \frac{p_p}{p_m} \quad \text{(approximately)} \quad (30.18).$$

We should realize that the terms *Kinetic* and *Potential Energies* were just old convenient terms, which are archaic and as dead as dinosaurs.

At present state of knowledge, we can surely state that any kind and any amount of energy can be described in terms of electrodynamics but not Newtonian mechanics. The lack of existence of kinetic and potential energies was obvious to Isaac Newton, who used only mechanical forces and momentum in his work.

From this point on, we must strictly adhere to the following rules:

- ➢ **Newtonian Mechanics describes inertial bodies by using mass, force, momentum, inertia, etc.**
- ➢ **Electrodynamics describes massless noninertial media.**
- ➢ **Inertial frame of references is defined by using the following tools of research:**

  - The tacitly assumed existence of a frame of reference (coordinate system) is where we could specify the position of an object under study from instant to instant by giving a certain number.
  - The means of measuring time (a clock) marks definite, equal intervals of time at which the position could be recorded.
  - Utilization of an "Inertial frame of references" is applicable for study of inertial and noninertial media because it considers not only Galilee's transformations that pertained to material bodies but also Lorentzian transformation that correlates influence of noninertial media as well.

- ➢ **Based on Fitzgerald's space contraction, the existence of "noninertial frames of references" of any kind is impossible.**

AFTERWORD

Back to my school years, my teacher Nathan Guberman, holding a white chalk in his hand, turned away from the blackboard and threw a question in my direction. "Does space really bend or turn under the influence of time?" He was speaking in a very quiet tone, like speaking to himself, with an inquisitive look in his eyes.

It was an unusual twist on his interpretation of General Relativity. Then and for many years afterward, I did not dare to look for the answer; I was afraid to crush some of my well-established beliefs. In 1998, during one of our meetings, Professor Bethe, with the same kind, inquisitive look in his eyes, asked if I believed in the Big Bang theory. I looked back and replied, "Definitely not." I knew the answer to his question, but I did not dare crush some of his well-settled beliefs.

In writing this book, I dare readers to do precisely that—challenge their outmost settled beliefs. My goal is to call attention to inconsistencies of well-accepted theoretical assumptions and to point out some alternative ways to describe the uniqueness of natural phenomena. Having tough times during the last decade to prove the discovery of the "Parker Effect," I have no illusions that my work would be met with open hands and hearts, neither do I expect immediate acceptance of the Dual Transformation and other theories described herein. Rather, knowing the present state of the scientific community, I foresee an uphill struggle and a lot of slandering.

I put my thoughts in writing because I do believe that sooner or later, the new generation of thinkers will defy the convenience of the conventional and adopt new ways of thinking that will lead to even more breathtaking discoveries.

New ways of thinking will encompass the thorough review of the mathematical apparatus of Quantum theories due to the abolishment of the outdated concepts of Kinetic and Potential energies.

The biggest, by far, will be the task to extend the theory of Dual transformation from the field of electrodynamics into the field of particle physics, describing the ways of the birth of particles. Hence, build a bridge between Newtonian mechanics and Maxwellian electrodynamics.

REFERENCES

Becket, Richard. 1964 –1982. *Electromagnetic Field and Interactions*. Dover Publications Inc.

Bondi, Hermann. 1961. Endeavour, Cambridge University Press.

Collins, Royal Eugene. 1999. *Mathematical Methods for Physicists and Engineers*. Dover Publications.

French, A. P. 1968. *Special Relativity*. W. W. Norton & Company.

Galilei, Galileo. 1953. *Dialogue on the Great World Systems*. University of Chicago Press.

Hey, Tony, and Patrick Walters. 1997. *Einstein's Mirror*. Cambridge University Press.

Joos, George. 1958–1986. *Theoretical Physics*. Dover Publications Inc.

Longair, M. S. 1984. Cambridge University Press.

Lorrain, Paul, Dale P. Corson, and Francois Lorrain. 1988. *Electromagnetic Fields and Waves*. W. H. Freeman and Company.

Mook, Delo E., and Thomas Vargish. 1987. *Inside Relativity*. Princeton University Press.

New Webster's Dictionary. 1993. Lexicon Publications Inc.

Pais, Abraham. 1982. *The Science and the Life of Albert Einstein*. Oxford University Press.

Parker, Alexander, and Val Parker. 2000. *Special Relativity and Navigation in Space*. "The Parker Effect and Contemporary Relativity." Paper presented at the American Physical Society Centennial Meeting Program.1998.

"What We Have Not Learned from the Troubles with the Hubble." *IEEE Aerospace and Electronic Systems Magazine*.

Standard Model of Fundamental Particles and Interactions. 1988–1999. Contemporary Physics Education. CPEP, Lawrence Berkeley National Laboratory, Berkeley.

Swartz, Clifford. *Phenomenal Physics*. The University of New York at Stony Brook.

Thorne, Kip S. 1994. *Black Holes and Time Warps*. W. W. Norton & Company.

The Encyclopedia Americana, International Edition.

Tipler, Paul A. *Physics*.

www.ingramcontent.com/pod-product-compliance
Lightning Source LLC
Chambersburg PA
CBHW030757180526
45163CB00003B/1064